Sustainable Energy

Sustainable Energy

An Annotated Bibliography

Namir Khan
Willem H. Vanderburg

with the collaboration of José Etcheverry,
Norman Freeman, Lynn Macfie, and
Esther Vanderburg

The Scarecrow Press, Inc.
Lanham, Maryland, and London
2001

SCARECROW PRESS, INC.

Published in the United States of America
by Scarecrow Press, Inc.
4720 Boston Way, Lanham, Maryland 20706
www.scarecrowpress.com

4 Pleydell Gardens, Folkestone
Kent CT20 2DN, England

British Library Cataloguing-in-Publication Information Available

Library of Congress Cataloging-in-Publication Data

Khan, Namir, 1955–
    Sustainable energy : an annotated bibliography / Namir Khan and Willem H. Vanderburg.
        p. cm.
    Includes bibliographical references and index.
    ISBN 0-8108-4044-8 (alk. paper)
        1. Energy development—Environmental aspects—Bibliography. 2. Sustainable
    development—Bibliography. I. Vanderburg, Willem H. II. Title.
Z5863.E54 K453 2001
[TD195.E54]
016.33379—dc21   2001020522

♾™ The paper used in this publication meets the minimum requirements of
American National Standard for Information Sciences—Permanence of
Paper for Printed Library Materials, ANSI/NISO Z39.48-1992.
Manufactured in the United States of America.

CONTENTS

PREFACE

This annotated bibliography introduces readers to preventive approaches for making the energy requirements of modern ways of life more sustainable. Since a society can neither create nor destroy energy, all energy is temporarily borrowed from the biosphere and returned to it in degraded form. Solar radiation is included since it is "prepared" by the ozone layer. The activities of a way of life of a society that involve energy flows may be represented as a network for which the biosphere acts as the ultimate source and sink. This network and therefore the energy basis for a modern way of life will be sustainable when the biosphere can perform these roles indefinitely.

Making the network of energy flows more sustainable should not involve transferring adverse consequences to other dimensions of sustainable development. Three primary dimensions may be distinguished. The metabolic dimension refers to making a way of life sustainable with respect to its dependence on the biosphere for all matter and energy. The habitat dimension is associated with the creation of a life-milieu that interposes itself between human life and the biosphere, of which cities are the principal physical component. The third dimension of sustainable development relates to human life and society, ensuring that a way of life does not deplete its "social capital" any more than its "natural capital." These dimensions of sustainability will be covered in companion volumes. In this larger context, the present bibliography regards the preventive approaches related to energy as a necessary but not sufficient subset of preventive approaches for improving the sustainability of contemporary civilization.

These bibliographies are the result of a twelve-year research project carried out at the Centre for Technology and Social Development University of Toronto. The point of departure was an examination of professional education affecting the engineering, management, and regulation of modern technology in the context of making modern ways of life more sustainable by reducing the harm done to human life, society, and the biosphere. During undergraduate education, future practitioners gain little knowledge of how technology affects these contexts and even less of how such understanding can be used in a negative feedback mode to adjust design and decision-making to prevent or greatly reduce harm. As a result, a great deal of design and decision-making occurring from the vantage-point of a particular discipline or professional specialty externalizes those consequences that fall beyond their boundaries. The result is that, in modern technology, problems tend to be first created and then remedied. This is needlessly costly and harmful.

Specialists tend to cope with this situation by implicitly and explicitly saying that they do the best they can in their area of competence and that those consequences of their actions and decisions that fall beyond these areas are best dealt with by others who have a competence there. We cannot help but sound a little defensive because we all know that our effectiveness and responsibility require both a frontier knowledge of our areas of competence and a context knowledge of those areas in which the consequences of our decisions and actions will fall. Only then will we be able to use this context knowledge to assess the consequences of our actions and decisions in order to adjust them to prevent or greatly reduce harm and, in the best of all worlds, ensure that the result will be technological and economic development instead of mere growth. At present, the "knowledge system" allows the consequences of the decisions and actions of one specialty to be dealt with only in an end-of-pipe or after-the-fact manner by specialists in whose areas these fall. It precludes dealing with the root of any problem. There is growing evidence that this is costly and ineffective because it frequently merely displaces symptoms from one area to another. In the areas of technological and economic growth, educating the practitioners involved in these enterprises as both specialists and generalists could make a fundamental difference in ensuring that any advances are not undermined by the problems they simultaneously create. This would lead modern civilization to adopt what I call preventive approaches in order to get us out of the labyrinth of the social and environmental implications of modern technology in which we are now lost because the "system" makes it almost impossible to get to the root of any situation or problem by means of specialized knowledge and practice.

The above situation is both cause and effect of the kinds of values that guide technological and economic growth. Performance values such as efficiency, productivity, cost-effectiveness, profitability, risk-benefit ratios, and GDP are ratios of desired outputs obtained from requisite inputs. For example, GDP measures the value of all goods and services produced by a society using its "social and natural capital." Hence, performance values give us no indication to what extent any technological or economic gain in output has partly or wholly been achieved by degrading the human, societal and biospheric contexts on which it depends. In other words, they give us no indication to what extent we are using our "social and natural capital" sustainably and to what extent we are "mining" them. For example, a new production system may lead to an increase in labor productivity and profitability and also produce more unhealthy workers, in turn affecting the fabric of their social relations and their communities. New materials may be particularly advantageous in terms of their technical and economic performance, but collectively such materials may undermine the health of ecosystems on which all life depends. In general, it may be said that technological and economic growth is characterized by gains occurring predominantly in the domain of performance values and problems in the domain of context values. The latter assess compatibility with the human, societal, and biospheric contexts.

As a consequence, the members of contemporary civilization individually and collectively "steer" technology, much like driving a car with the windows covered, leaving the driver to concentrate on the performance of the vehicle as indicated by the instruments on the dashboard. Each time a scream is heard or a bump is felt, we frantically attempt to rip holes in the covers over the windows to see what is happening, but this is an after-the-fact reaction. After two centuries of technological and economic growth, there are many small holes in the window covers but we still do not have a clear view of how technology is related to everything else on the road of human history.

Following this diagnosis of the present situation, the research program undertook an exhaustive comparison of conventional with state-of-the-art methods and approaches for dealing with the social and environmental implications of technology. The former deal with any technological situation or problem in two stages. The first seeks to obtain the highest possible desired outputs from the requisite inputs. Success is measured in terms of performance values. A second stage deals with those undesired outputs that are prohibited or constrained by laws and regulations. The state-of-the-art methods collapse these two stages into one, at least with respect to one dimension of sustainable development. Pollution prevention is paradigmatic. It goes back to the root of the problem and asks why a particular product or process produces pollutants in the first place. How can the material inputs and/or the process be modified to avoid producing these pollutants? The possibilities do not end with pollution prevention. There are other approaches, and it is only after all these fail that end-of-pipe abatement should be attempted. This led to the formulation of the concept of preventive approaches, which incorporate into design and decision-making a proactive consideration of the implications to "steer" technology to avoid collisions with its contexts. The research findings led to the development of a new conceptual framework and strategy aimed at converting technological and economic growth into development that would gradually become more sustainable. It was put into practice in a set of courses and a certificate program aimed at educating engineers with a difference. The present bibliography brings together that portion of the literature that can make a significant contribution to preventive approaches in the area of energy within the metabolic dimension of sustainable development.

The references are organized according to the following plan.

Diagnosis and Level of Analysis

For the purpose of the diagnosis, analysis, and implementation of preventive approaches, the activities that make up the way of life of a society may be aggregated as a whole, as belonging to a "system" involving those activities related to a specific material or product or to a service rendered by means of that material or product, or as related to a particular phase in the technological cycle of a material or product. The resulting network of activities is not sustainable because it exceeds the capacity of the biosphere to act as the ultimate source and sink on a continued basis.

A. Representation of Network

The energy requirements of the way of life of a society may be represented as a network of flows of energy that form the inputs and outputs of the human activities represented as the nodes. Since a society can neither create nor destroy energy, all inputs and outputs into and out of this network derive from the biosphere as the ultimate source and sink of all energy flows. This network partly overlaps and is closely related to the ones representing the flows of materials in a society. The energy network is built up from *fuel cycles* involving the acquisition of the inputs into the cycle from the biosphere, their refinement, preparation and distribution for use. These fuels are directly converted into energy and wastes in the provision of a service to society or converted into *energy carriers* that can be distributed for subsequent consumption. Energy may also be captured directly from the biosphere, as is the case in solar, wind, and geothermal energy. Since all energy transformations are irreversible, energy flows through the network in linear chains from which the maximum useful energy must be extracted as in co-generation, for example. Limited representations of the energy network include: *input-output methods* representing the flows of fuels in monetary terms; *mass-energy balances* of an economy converting financial flows into those of specific forms of energy; and *life-cycle analysis inventory stages* representing the flows of energy in a materials or product system.

B. Restructuring the Network for Greater Sustainability

1. *Energy Supply*

a. Non-renewable energy sources: The last two decades have seen considerable advances in making energy supply from conventional resources much more efficient, thereby reducing the required inputs into the network. The efficiency of coal, oil, and gas-fired energy systems has been increasing every year, and when combined with other strategies such as de-centralization, cogeneration (combined heat and power generation), appropriate matching of energy source with energy service, as well as district heating and cooling systems, etc., such energy systems can be much more preventive than conventional, large scale, centralized systems.

b. Renewable energy sources: Supply strategies that concentrate exclusively on renewable energy sources aim at making the network indefinitely sustainable.

2. *Energy End-Use*

a. Demand side management is an important route to increasing energy efficiency as well as reducing demand through other measures. *Integrated resource planning* seeks to make the network more sustainable by comparing supply side and demand side options. The latter is regarded as negatively producing energy. Some versions of integrated resource planning emphasize the amount of energy saved as the most important resource (as opposed to those that advocate it as a resource equivalent to others and to be judged only in terms of cost).

b. Increasing the efficiency of energy end-use makes the network perform the same services using less energy. *Energy analysis, energy end-use analysis,* and *energy accounting* make important contributions to energy efficiency.

3. *Energy Policy for Sustainability*

The findings of the above two categories should be translated into *energy policy* that explicitly incorporates sustainability issues.

The following categories of preventive approaches are usually dealt with in the literature on industrial ecology, although even there the energy implications are generally poorly developed. This literature has therefore been covered in the bibliography on industrial ecology.

C. Initiatives Related to the Entire Technological Cycle of a Specific Material or Product or the Service Rendered by Them:

Design for Environment (DFE): An inventory of the implications for sustainable development of a particular material, product, system, or service during its entire technological cycle informs design and decision-making to reduce negative effects as much as possible. This includes the impacts of energy conversions.

Product stewardship: The emphasis is shifted from the sale of fuels and energy carriers to the supply of the services they render, whereby a utility or corporation accepts responsibility or is made accountable by law for the entire technological cycle of a material or product, including energy requirements or services involving materials or products.

Partial product or service stewardship: A utility or corporation takes back materials and products (that perform a service through the conversion of energy) at the end of their use-phase.

Energy labeling: Decisions are informed by a description of the environmental characteristics of the energy transformations directly and indirectly involved in a product system or a service over its entire technological cycle.

D. Initiatives Related to a Particular Phase in the Technological Cycle of a Specific Material, Product, or Service

DFE should ensure the greatest possible use of renewable resources and energy end-use efficiency for each phase in the technological cycle of a material, product, or service.

It is hoped that this structure will facilitate specialists in becoming generalists in those domains where the consequences of their actions are most important, help policy-makers and politicians come to grips with the broader implications of "steering" technology, as well as help members of the public gain a better all-around understanding of the complexity of the many issues that surround technological and economic growth. A better understanding of the ecology of technology and the ecologies of individual technologies is essential to create a humane and sustainable future. The bibliography is therefore not intended to be encyclopaedic. It is designed to support the development of preventive approaches, and the selections have been made with this general purpose in mind. In terms of the division of labor between co-authors, the overall direction of the project and the development of the conceptual framework have been my primary responsibility, while the gathering and abstracting of the particular contents of this bibliography have been under the direction of my co-author. The collaborators prepared significant numbers of abstracts and assisted with the editing and organization of the bibliography.

Willem H. Vanderburg
Toronto, January 2001

SUSTAINABLE ENERGY BIBLIOGRAPHY

Representation of Network

This section of the bibliography deals with writings that look at energy in a multiplicity of ways to reveal the various social, political, economic, and environmental dimensions in which its supply and use are embedded. Most comprehensive discussions of energy that incorporate sustainability issues (whether explicitly or implicitly) will be found here. Many of these works also make important contributions to subsequent sections of this bibliography. Peter Chapman's *Fuel's Paradise*, for instance, is not only a landmark account of how energy is implicated in all aspects of our life, highlighting the various strands of the "network," it also contributes to energy analysis, an analysis of the environmental impact of energy production, energy scenario construction, and end-use efficiency.

1. Ausubel, Jesse H., and H. Dale Langford, eds. *Technological Trajectories and the Human Environment*. Washington, D.C.: National Academy Press, 1997.

Keywords: economics, efficiency, environmental

Three articles that relate to energy strategies for preventive engineering may be found here. These are: "Freeing Energy From Carbon," by Nebojsa Nakicenovic; "Life-Styles and the Environment: The Case of Energy," by Lee Schipper; and "Electron: Electrical Systems in Retrospect and Prospect," by Jesse H. Ausubel and Cesare Marchetti. Nakicenovic discusses the global decrease in energy requirements per unit of economic output and the adoption of various energy efficiency measures through which this lowering of energy intensity was achieved. She also examines decreases in carbon dioxide emissions from fossil fuel based energy systems stemming from various technological innovations. Lee Schipper uses the case of energy to indicate the overall level of environmental disruption caused by different demands for services. The strength of Schipper's argument lies in his focus on the relationship between lifestyles and the level of demand on energy services. Achieving sustainability in the arena of energy may therefore involve some change in an essentially consumerist culture. Ausubel and Marchetti examine electricity over time, from its first identification in amber by the Greeks, to steam generators in the modern era, and to the foundation and consolidation of electric companies in the United States. They also give a comprehensive analysis of the following aspects of the electricity sector in the United States: the capacity of power lines; installed electric generating capacity; the rate and total consumption of electricity; percentage of homes with electric service; improvements in motor efficiency; and other efficiency improvements under development.

2. Bhagavan, M. R., and Stephen Karekezi, eds. *Energy for Rural Development*. London: Zed Books, 1992.

Keywords: alternative energy, biomass, case studies, economics, renewable, supply and demand

This is a comprehensive analysis of the use of new and renewable sources of energy in integrated rural development. The strength of the book derives from the fact that the various authors realize that there is no single magic bullet to development (and certainly not through the mere provision of energy), nor are there any universal solutions; development solutions must work on a number of fronts simultaneously so that benefits that accrue on one front are not undercut by setbacks in other areas. Solutions also have to be context-sensitive; without a full awareness of prevailing cultural, economic, and political forces, development efforts are likely to fail. The editors have not shied away from including accounts of failed projects. There is much to be learned from these failures. A case study of one such project in the Philippines, involving a wood-fired thermal power plant, is a classic example of what must not be done in a development project and how the best-intentioned of efforts can lead to consequences where the local people involved are worse off after development efforts than they were before. Other case studies go on to demonstrate how carefully planned initiatives in the energy arena can effectively achieve a wide range of development goals. There are also succinct analyses of various energy sources, from biogas to agricultural residues, and a good discussion of their pros and cons. The remainder of the book contains descriptions and analyses of various renewable energy projects completed or under way in Africa as a whole, China, India, Sri Lanka, Mexico and the Caribbean, parts of the former Soviet Union, and Eastern Europe.

3.   Blanchard, Odile. "Energy Consumption and Modes of Industrialization." *Energy Policy* (December 1992): 1174-85.

Keywords: case studies, consumption, efficiency

This is a comparison of energy consumption in South Korea, India, Brazil, and Mexico. These countries are looked at individually because they represent various stages of development, and together, they are responsible for more than a third of all the energy consumed in developing countries (except for China). The difference in energy consumption in these countries can be attributed to three factors: "the availability of national energy sources," "industrialization mode as a major determinant of final energy consumption dynamics," and "private consumption patterns and energy consumption dynamics." Blanchard analyzes a number of statistics on population and GDP, energy dependence, energy consumption per capita, energy consumption by source (solid mineral fuels, oil, gas, primary electricity, and non-conventional energies), and primary energy consumption by industry, as well as the transport, residential, tertiary, and agricultural sectors. Efficiency is seen as a key element in reducing energy consumption.

4.   Budnitz, Robert J., and John P. Holdren. "Social and Environmental Costs of Energy Systems." *Annual Review of Energy* 1 (1976): 553-80.

Keywords: economics, environmental

The environmental and social costs of various technological policies and activities are often neglected. In particular, new energy technologies are examined primarily in terms of economic cost, and are therefore susceptible to social and environmental "mistakes." In examining the wide spectrum of these environmental and social costs (i.e., occupational safety, public health, economic productivity, environmental diversity, and social stability), this article discusses impact analysis both in policy and action. It reviews the available methods for such evaluations as they relate to energy technologies. Numerous examples are used and the key concepts are well defined.

5.   Byrne, John, and Daniel Rich, eds. *Energy and Cities.* Energy Policy Studies, Vol. 2. New Brunswick, N.J.: Transaction Books, 1985.

Keywords: cities, conservation, economics, efficiency, environmental

     The types and quantities of energy used by cities have played a fundamental role in their development as well as indicating the path that they might follow in the future. Each of the five articles that make up this book addresses issues raised by the changing relationship between cities and energy. The main debate among them is whether cities will "re-concentrate" as a result of a coming energy crisis, or "de-concentrate" in terms of population and space. Other issues discussed are: "the role of cities as energy resource allocation systems"; barriers to saving energy in the urban built environment; the international economic and political power that energy can have over communities; the need for cities to change their role from "consumers of goods to producers of wealth"; hard versus soft energy paths; methods of energy production, including decentralized energy production; the impacts that changing forms and uses of energy will have on society and the social costs involved in these transformations. A select bibliography on energy and cities enhances the value of this collection.

6.   Byrne, John, and Daniel Rich, eds. *Planning for Changing Energy Conditions.* Energy Policy Studies, Vol. 4. New Brunswick, N.J.: Transaction Books, 1988.

Keywords: case studies, conservation, developing countries, economics, efficiency, electricity, environmental, planning, policy, transportation

     The challenge posed to planning by changes in energy issues is examined from a multidisciplinary perspective. The analyses look at planning considerations from the local to the national and international levels. The main argument of the first chapter is that sustainability concerns have made it imperative that planning systems make energy efficiency their primary aim. The second chapter analyzes electric power planning in the third world in terms of organization and goals. Chapter 3 examines the various unanswered and perhaps unanswerable questions involved in predicting the hydro-geological stability of burial sites for radioactive waste. Chapter 4 discusses coal development in relation to U.S. land policy. Chapter 5 analyzes energy policy and planning in the UK in relation to coal and regional development. Chapter 6 covers theoretical and simulation results of the effects of oil price changes on urban structures. Chapter 7 focuses on the prospects and potential of community energy planning, and chapter 8 explores the future of energy and planning.

7.   Cartledge, Bryan. *Energy and the Environment.* Oxford: Oxford University Press, 1993.

Keywords: alternative energy, environment, fuels, nuclear, sustainability

The positive and negative aspects of various forms of energy generation are examined using an approach that combines comparative statistics with philosophy. Cartledge begins by giving an overview of the different problems that developed and developing countries face in terms of energy and the environment, before moving on to a discussion of sustainable development, noting that this term is being bandied about much too lightly. The author asserts that if something is said to be sustainable, that should mean that it should continue indefinitely; this does not mean that true sustainable development is impossible, only that it requires intelligent decisions. By way of solutions he suggests ways in which we can achieve cleaner power from fossil fuels by increasing efficiencies in the steam cycle, using combined cycle gas turbines, bringing on-line advanced coal conversion technologies, and speeding up the deployment of other technologies in development. The gas industry is examined in some detail and the environmental impact of gas fired technologies is described. Alternative energy sources in Europe are discussed next, along with mention of available policy options, their implementation, and the barriers they face. Various strategies are suggested to overcome these barriers. Environmental concerns related to predominant forms of energy generation in Central and Eastern Europe are described in detail, and the legacy of Chernobyl on the prospects for energy generation in the former Soviet Union is analyzed. Cartledge concludes with a provocative commentary on nuclear power not being as potentially harmful to human populations and to the environment as people seem to think. Other than this sudden lapse in judgement, the book makes many suggestions that are consistently preventive.

8.   Chapman, Peter. *Fuel's Paradise: Energy Options for Britain*. London: Cox & Wyman, 1975.

Keywords: consumption, energy analysis, economics, environment, fuels, policy

Chapman was the first to popularize the field of energy analysis with his fabulous tale of "The Isle of Erg" (where energy units function as currency instead of money), with which *Fuel's Paradise* begins. He was also among the first to use energy analysis as a comprehensive technique to determine how fuels are really used in the United Kingdom, and to calculate possible variations in fuel demand associated with different lifestyles. He analyzed various factors, such as detrimental climatic effects and resource depletion, that can reduce the use of fuel. This analysis provides the background to the book's main theme, which is an exploration of available energy policies. Three policy options are discussed in detail, representing a wide spectrum of choice. The first option is a "business as usual" scenario, where current trends which make our economic system function are assumed to continue. The second option is based on a "technical fix" approach, while the third is based on a "low-growth" scenario. The author notes that each of these three options is linked to a unique set of problems. In the first case, environmental sustainability is adversely affected, and should global warming become a reality, the consequences for human life would be catastrophic. Population growth and development needs similarly make the low-growth scenario unacceptable, since it means a decline in lifestyles when precisely the opposite is needed. The technical-fix scenario is seen as the best available option even though it still has some negative resource implications and may be economically too expensive to implement.

9.   Cullingworth, J. Barry, ed. *Energy, Land and Public Policy*. New Brunswick, N.J.:

Transaction Books, 1990.

Keywords: case studies, economics, environment, planning, sustainability

One of the main ideas uniting this collection of papers is that different energy supply systems significantly influence land use patterns, and hence the structure and livability of cities. The range of topics is quite vast, from a historical analysis of the relationship between energy supply technologies of the nineteenth century and the structure of cities, to an analysis of how energy demand and conservation issues have affected the spatial structure of modern cities. Other essays examine the problems associated with existing nuclear waste sites in the United States; the regional implications of energy price fluctuations; and trends in the energy intensity of the manufacturing sector in the United States and Canada. A final essay makes useful recommendations on how the energy sector can be made more sustainable.

10. Curran, Samuel L., and John S. Curran. *Energy and Human Needs.* Edinburgh: Scottish Academic Press, 1979.

Keywords: alternative energy, electricity, end-use, environment, nuclear, policy, transportation

Designed as a textbook, this is a comprehensive overview of energy issues, including discussions of social and environmental implications of energy generation and use. Fundamental thermodynamic concepts are defined, and then essential concepts underlying energy flows and energy conversion are discussed before the reader is introduced to a more detailed analysis of energy sources, modes of generation, and end-uses. A discrete section of the book deals entirely with the environmental impact of energy generation and use. Particularly useful here is a comparison of environmental damage resulting from sources of energy generation.

Subsequent sections give a detailed analysis of nuclear energy, its economics, and the safety concerns associated with it. Details of electricity generation and its end-uses are presented, using energy analysis methodology. This section is very useful in outlining all the inappropriate end-uses of this high quality energy source, such as heating, and suggests alternative energy sources for these uses that are more appropriate. A concluding section is devoted to global issues that range from international security and nuclear hazards, to future policy options with regard to environmental protection.

11. Eberhard, Anton, and Clive Van Horen. *Poverty and Power: Energy and the South African State.* London: UCT Press, 1995.

Keywords: case studies, efficiency, planning, poverty alleviation

Poverty has traditionally been accompanied by a lack of energy resources, and this continues to be true today. In reminding us of the importance of equity as a development goal, and the crucial role of the state in its achievement, the author discusses ways to widen access to improved energy services for people. South Africa is used as a case study in exploring this theme. Methodological approaches and analytical tools that facilitate coherent, rational, and efficient state involvement in the energy sector are also described. The lessons learned here are applicable to other countries as well. The following topics are covered: the role of the state in achieving greater equity; the changing context for energy planning in South Africa; household energy and poverty; accelerated electrification as the core of South Africa's household energy policy; policies for fuelwood, hydrocarbons, and energy efficiency; and a new system of governance for the energy sector.

12. Elliott, David. *Energy, Society and the Environment.* Environment and Society. New York: Routledge, 1997.

Keywords: economics, efficiency, environment, fuels, nuclear, renewable, sustainability

The way energy use interacts with society and the environment is examined in this book. The author concentrates on alternatives based on possible policy options and new technologies. Part 1 examines the following issues: environmental problems that result from energy use; energy growth since the Industrial Revolution and growing environmental problems; sustainable development as an alternative approach; and sustainable energy technologies. In part 2, a full range of renewable energy technologies are analyzed, which may aid in creating an environmentally sustainable energy future. This section also describes "green" consumer products and domestic energy conservation techniques; more efficient uses for fossil fuels; and nuclear power. Problems of feasibility are addressed in part 3, in terms of issues related to the development and implementation of sustainable energy options. The focus of this section is on technical, economic, and strategic issues. The author also analyzes institutional difficulties and public reluctance in accepting new technologies. In the concluding section, sustainability is analyzed in terms of the limits of technical fixes, whether these fixes will encourage economic growth, and what other changes may be necessary for achieving sustainability.

13. Energy Research Group. *Energy Research: Directions and Issues for Developing Countries.* Ottawa: International Development Research Centre/United Nations University, 1986.

Keywords: developing countries, environmental

This report surveys energy research and suggests priorities for developing countries. The report is based on three premises: energy research must be related to research on the entire economy and society; energy sources must be studied in the context of demand for them; and energy saving is as important as energy production. It begins by analyzing the setting for research in developing countries and the role played by governments, research institutions, and international funding agencies. Several approaches to energy-demand management and conservation are outlined, and these are illustrated with examples of implementation opportunities for certain common types of equipment. Research priorities for various energy sources such as thermal energy, motive power, and electricity are also identified. While there are many pervasive environmental effects emanating from energy production and use, the report focuses on four environmental issues related to energy: deforestation, desertification, climate change, and acid rain.

14. Flavin, Christopher, and Nicholas Lenssen. *Power Surge: Guide to the Coming En-*

ergy Revolution. New York: Norton, 1994.

Keywords: solar, photovoltaics, policy, transportation, nuclear, sustainability

The authors maintain that an energy revolution is beginning. This revolution is given impetus by the numerous detrimental effects caused by our reliance on fossil fuels, and is stimulated by rapid technological innovations in energy generation and conservation. The emergence of this energy revolution is not yet accepted by most mainstream energy organizations and planners, which see the future as just a continuation of the present with only minor adaptations. The authors forecast that the emerging alternative energy path will be based on the widespread adoption of energy efficiency strategies: the conversion of coal and nuclear plants into efficient gas turbines as a transitional step towards a future dominated by hydrogen fuel generated by renewable means; widespread decentralized energy generation facilitated by wind turbines, solar thermal applications, and photovoltaic cells; and innovations in automobile and building design. The authors provide numerous examples from different nations and analyze different forecast figures to substantiate their arguments. Policy suggestions to usher in the coming energy revolution are also offered.

15. Foley, Gerald. *The Energy Question*. 4th ed. London: Penguin Books, 1992.

Keywords: developing countries, energy analysis, renewable

A significant segment of the human race is becoming aware that current lifestyles are jeopardizing our own survival. In terms of energy, there is a widespread concern that the predominant forms of energy production and use are causing inevitable and detrimental changes to the planet's climate. However, for various reasons, renewable forms of energy generation have not yet become a common part of the mainstream. This book is intended to provide background and current information on energy issues and to guide the reader on the most significant concerns among them. The book is divided into four parts covering linkages among energy, society, and resources, as well as energy in the developing world, and choices for a different energy future.

16. Foley, Gerald. "Rural Electrification in the Developing World." *Energy Policy* (February 1992): 145-52.

Keywords: developing countries, economics, electricity

This article disputes the notion that rural electrification directly enhances development. The author suggests that demand for electricity is a consequence, rather than a cause, of economic development. His arguments raise challenging questions about the costs and benefits related to the allocation of scarce resources in developing nations. Some of the issues analyzed in this article are: the role of subsidies and tariffs; the selection of rural areas for electrification; choice of supply technologies: conventional versus renewable alternatives; and obstacles related to solar electrification projects.

17. Goldemberg, Jose, Thomas B. Johansson, Amulya K. N. Reddy, and Robert H. Williams. *Energy for a Sustainable World*. Washington, D.C.: World Resources Institute,

1987.

This book is a shorter version of *Energy for a Sustainable World* published by Wiley-Eastern, New Delhi, India. For an overview, see below.

18. Goldemberg, Jose, Thomas B. Johansson, Amulya K. N. Reddy, and Robert H. Williams. *Energy for a Sustainable World.* New York: John Wiley & Sons, 1988.

Keywords: policy, case studies, economics, energy analysis, developing countries, sustainability, end-use, poverty

If all available energy efficiency measures were put into place and policy efforts were redirected toward the satisfaction of human needs rather than increasing the supply of energy, the entire world would be able to achieve a standard of living comparable to that of Western Europe in the 1970s, according to the authors. This startling claim is substantiated through an alternative approach to conventional energy analysis. This approach does not consider economic growth as the single objective of energy analysis. It also maintains that equating higher energy use to human welfare is a misguided strategy that results in multiple detrimental effects. The authors propose an alternative approach to conventional energy planning that is based on the incorporation of broad societal goals such as equity, self-reliance, economic efficiency, environmental harmony, long term viability, and peace. In order to incorporate these societal goals they explore an "end-use methodology" which aims to understand the human needs that energy serves and explores the most effective ways to ensure that energy resources satisfy these needs. It is maintained that the implementation of the energy strategy presented here will result in a more equitable, economically viable, and environmentally sound world. The volume is divided into six chapters that analyze in great detail the following topics: an end-use oriented energy strategy for industrialized countries; energy strategies for developing countries; global energy demand and supply; policies for implementing energy strategies for a sustainable world; the political economy of end-use strategies; and renewable energy resources. This is an inspirational, strong, and exhaustive argument that provides scenarios to achieve a sustainable future in which all benefit. As the authors suggest: "contrary to widely held beliefs, the future for energy is very much more a matter of choice than of destiny."

19. Goldemberg, Jose. *Energy, Environment and Development.* London: Earthscan, 1996.

Keywords: policy, economics, sustainability, environmental impacts, energy analysis

Several environmental problems are related to energy issues. The purpose of this book is to serve as an introductory text that explains the connections between energy and environment. Its focus is on the type and amount of energy consumed by different groups in society. This so-called "dis-aggregative" approach aims to differentiate among local, regional, and global environmental impacts in order to facilitate understanding of multi-level environmental protection measures.

The first chapters provide the conceptual groundwork necessary to understand energy, economic, and development linkages. Subsequent chapters present the causes of, and potential solutions for, energy-related problems that cause environmental degradation. These chapters are complemented by an analysis of global energy trends that serve as a background to a policy discussion on sustainable energy development.

20. Gore, Al. "Environmental Technologies for a Sustainable Future." *Environmental Protection Agency Journal* 20, 3 & 4 (Fall 1994): 6-8.

Keywords: developing countries, economics, environmental, sustainability

Since our environment and our economy are mutually dependent, we need to make responsible environmental policies in order to provide lasting economic opportunity. We must invest in and develop sustainable technologies—those that encourage economic growth and protect the environment at the same time. The goal is to maximize the protection of public health and the environment, while minimizing costs. New ways are being sought to manage whole ecosystems, rather than small, disparate pieces of natural habitat. A new program called the Environmental Technology Initiative (ETI), headed by the EPA, was designed to "spur the development and marketing of innovative technologies throughout the economy." The article also looks at how the EPA is providing technical advice to environmental companies in promoting environmentally safe technologies in developing nations and throughout the world. It ends with two small sections that examine both environmental technology and the ETI.

21. Graham, Stephen. *Energy Research in Developing Countries*. Ottawa: International Development Research Centre, 1994.

Keywords: alternative energy, case studies, developing countries, economics, fuels

Although energy issues have been widely discussed for quite some time, concerns regarding the use of energy have shifted from fear of shortages to concerns about excessive use and pollution. This book is a general summary of the fourteen volumes compiled by the Energy Research Group (ERG) on energy in developing countries. It covers the analysis, management, and conservation of energy demand; energy economics; electricity; oil and gas; alternative liquid fuels; nonconventional energy; bioenergy; human energy; energy modeling, information systems, research and development of energy planning; energy in Africa; energy in Asia; energy in China; and energy in Latin America.

22. Grubb, Michael. *Renewable Energy Strategies for Europe, Volume I: Foundations and Context*. London: Earthscan, 1996.

Keywords: policy, sustainability, case studies, energy analysis

This work is part of a four-volume collection that deals with the current and potential role of renewable energy in Europe. This first volume describes the basis for renewable energy policy in Europe. It also analyzes in detail the intellectual and political driving forces behind renewable energy initiatives; energy resources and projections; socioeconomic links; energy-environment connections; and the historical and international dimensions of various renewable energy projects. The interrelated nature of renewable energy issues and their political relevance to sustainability and structural policies in Europe are emphasized. This analysis aims to illustrate the ambiguous results that policy-making can engender in different national settings. The main goal of this work is to enhance the effectiveness of renewable energy efforts and policies.

23. Hill, Robert, Phil O'Keefe, and Colin Snape. *The Future of Energy Use*. London: Earthscan, 1995.

Keywords: economics, environmental, energy analysis, supply and demand, renewable

This is a concise yet exhaustive review of several forms of energy generation and use. The authors thoroughly analyze the economic issues, environmental impacts, and social costs related to conventional and renewable energy sources and provide estimates about their future supply and development potential. Several energy planning issues are discussed, including agro-forestry and fuel-wood programs in Africa, the relationship of energy to markets, and issues of equity. The role of end-use analysis is illustrated by references to energy use in transportation and buildings. The volume explores different ways to incorporate environmental and social costs into energy planning and accounting, and the development of policies to achieve a more sustainable energy future.

24. Hohmeyer, Olav. *Social Costs of Energy Consumption*. Berlin: Springer-Verlag, 1988.

Keywords: case studies, consumption, economics, policy, renewable, solar, wind

Hohmeyer compares and analyzes the costs and benefits of electricity production by means of fossil fuels, nuclear energy, wind energy, and photovoltaics in Germany during the 1980s. Although most of the figures are pertinent only to the Federal Republic of Germany, and are therefore chronologically and historically bound, the general approach used here could be very useful as a benchmark for the development of similar studies. The author addresses the effects that these forms of energy generation have on the environment, employment, the depletion of natural resources, and public subsidies. He notes that if these effects are not considered, the use of renewable energy sources will not compare favorably with the use of conventional energy sources. Ironically, addressing these effects can result in a distortion of energy markets. However, the author notes that remedial market corrections could be made through policy measures implemented by governments. Several prerequisites to this form of government action are identified; for example, the existence of accurate knowledge regarding the extent of the external effects involved. The study also analyzes several factors needed to provide a basis for government action that will enhance the use of wind and solar energy.

25. Hohmeyer, Olav, and R. L. Ottinger, eds. *Social Costs of Energy: Present Status and Future Trends*. Berlin: Springer-Verlag, 1994.

Keywords: economics, environment, sustainability, case studies

This is a collection of papers presented at the International Conference on the Social Costs of Energy held in Racine, Wisconsin, in September 1992. They examine the wide spectrum of effects related to energy production and use, and are an aid in charting a new course towards a more sustainable and equitable future. The volume is divided into four subject areas covering the assessment of social costs and the various ethical considerations that should be considered a part of energy costs; empirical estimates of social costs ranging from energy supply to demand-side technologies; instruments and approaches for the internalization of social costs; and the relationships of social costs to sustainable development. Most of the papers present case studies from Europe and the United States.

26. Hollander, Jack M., ed. *The Energy-Environment Connection*. Washington, D.C.: Island Press, 1992.

Keywords: bio-mass, economics, efficiency, environment, fuels, nuclear, policy, pollution, sustainability

As the title suggests, this group of essays address the links between energy and the environment. The main goal is to identify possible solutions that would help create a more livable and sustainable world. There are three sections: Part 1 examines the environmental impact of the use of different forms of energy, such as nuclear power, fossil fuels, and bio-mass. The resulting effects of air pollution, acid rain, oil spills, nuclear leaks, as well as damage done to the quality of water and land are analyzed. Part 2 discusses some of the environmental and economic benefits of energy efficiency that can be implemented in buildings, transportation, and manufacturing. Part 3 covers the economics and ethics of protecting the environment; working towards a sustainable world; and policy issues such as an appropriate energy agenda for the 1990s.

27. Irwin, Nancy. *A New Prosperity: Building a Sustainable Energy Future*. Boston: Brick House Publishing, 1981.

Keywords: efficiency, electricity, energy efficient buildings, economics, fuels, solar, sustainability, transportation

Continued economic growth will necessitate increasing efficiency in energy use because of the expense of enlarging the supply of fuels in the future. Efficiency measures are seen as an economic opportunity, offering a "new prosperity," rather than an economic cost as conventional wisdom suggests. Four areas in which economic benefits may be realized through energy efficiency measures are studied closely: buildings, industry, transportation, and utilities. Irwin begins by examining ways in which new residential and commercial buildings can be built using only about a quarter of the energy required for heating and cooling for a typical unit in the United States today, through means such as better insulation, tight construction, storm windows, day-lighting, and efficient furnaces and air-conditioning equipment. She then discusses transportation and fuel efficiency. The increased emphasis on small, efficient engines, policies for increased performance standards, and taxes are examined. Alternative fuels and the unfortunate shift of freight from rail to truck that is presently occurring are also discussed. Irwin's attention then shifts to energy supply, and she gives a good description of the uncertainties about future energy demands that complicate the investment strategies of gas and electric utilities. Solar electric generating devices are also looked at to determine whether they are an economically viable substitute for existing oil and gas burning technology.

28. Kaijser, Arne. "Redirecting Power: Swedish Nuclear Power Policies in Historical Perspective." *Annual Review of Energy and the Environment* 17 (1992): 437-62.

Keywords: case studies, energy analysis, nuclear

In early 1980 the citizens of Sweden decided, through a national referendum, to phase out all their nuclear power plants by the year 2010, and this historic decision was soon ratified by the Swedish Parliament. The author notes that this decision becomes even more remarkable when we consider the fact that Sweden was then, and still is today, one of the largest per capita users of nuclear power. This article discusses the implications of the Swedish shift from high reliance on nuclear power to nuclear phase-out. Throughout this analysis, a sociotechnical approach is taken, where the energy system is understood to consist not only of technical components, but also of the people and organizations that build, operate, and use these facilities. The article studies all aspects and stages of this monumental proposal, from the evolution of the Swedish nuclear age to the current transitional stage.

29. Kainlauri, Eino, Allan Johansson, Ilmari Kurki-Suonio, and Mildred Geshwiler, eds. *Energy and Environment.* Atlanta, Ga.: American Society of Heating, Refrigerating and Air-Conditioning Engineers, 1991.

Keywords: alternative energy, conservation, economics, environment, planning, pollution

This is the proceedings of the International Symposium on Energy and Environment that took place in Finland in 1991. The focus is on the effects of energy production on the environment. It also analyzes the energy economy and the relationships of regional and urban planning to power systems. The various sessions provide valuable information on the global carbon cycle and the different ways in which carbon may be sequestered to mitigate against global warming; climate modeling; energy generation and the greenhouse effect; energy generation and the production of other pollutants; alternative energy technologies; energy systems and regional and urban planning; and demand-side management. The session devoted exclusively to the energy economy is particularly useful in illustrating how energy is implicated in the web of all social interactions. The sessions on technology transfer and future developments in energy technology enhance the value of this collection.

30. Karekezi, Stephen, and Gordon A. Mackenzie, eds. *Energy Options for Africa: Environmentally Sustainable Alternatives.* London: Zed Books, 1993.

Keywords: case studies, conservation, efficiency, environmental, photovoltaics, planning, policy, solar, sustainability

This book is based on papers presented at an African Energy Experts Meeting in 1992. Their discussions focused on ecologically sound alternatives that can enhance sustainable development in Africa. The central issues of this book are energy use in sub-Saharan Africa, its effect on the environment, and the need for change in terms of the approach used by African energy planners. The following topics are covered in detail:

environmentally sound energy strategies for sub-Saharan Africa; innovative policy instruments and institutional reform; training and human resource development for the energy sector; maintaining Ethiopia's power sector; mobilizing local financial resources; the role of the African Development Bank in energy development; a regional perspective of energy efficiency in industry; energy efficiency and the transportation sector; national industrial energy conservation; photovoltaics and solar water heaters; and the tools and methods of energy and environmental planning in the 1990s.

31. Kleindorfer, Paul R., Howard C. Kunreuther, and David S. Hong, eds. *Energy, Environment and the Economy: Asian Perspectives.* Cheltenham, England: Edward Elgar, 1996.

Keywords: case studies, economics, environment, planning, pollution prevention

This volume contains several papers presented at the conference on Energy, Environment, and the Economy held in Taipei, Taiwan in 1994. The conference provided a forum for experts from Asia, Europe, and North America to discuss alternatives for balancing the issues of energy planning, environmental management, and economic development. The introductory articles provide a useful overview of current energy policy in Asian countries and the various environmental concerns that demand changes to this policy. The move toward sustainability is hampered by a demand for economic growth to raise the abysmal standard of living in some of the countries represented. Other articles suggest policies through which such growth may be brought about while reducing the energy intensity of the economy at the same time. Various other strategies are described as well, ranging from tradable emission permits to attracting global capital to invest in efficiency measures. Particularly useful are articles describing community-based and managed energy systems and the policies through which these inherently more sustainable systems may be brought about. A discussion of quality of life indicators more appropriate for an Asian setting is also well articulated and would benefit policy-making in the area of distributional decisions. A discrete section on lessons to be learned from the European experience in energy generation as well as associated efforts to fight pollution and dispose of hazardous waste is also useful. Case studies from Japan, South Korea, Taiwan, Indonesia, and the Philippines increase the value of this book.

32. Kraushaar, Jack J., and Robert A. Ristinen. *Energy and Problems of a Technical Society.* 2nd ed. New York: John Wiley & Sons, 1993.

Keywords: case studies, conservation, fuels, generation, nuclear, pollution, solar, supply and demand, wind

This is a useful introductory volume to the field of energy. It analyzes different forms of generation, supply and use, and the consequences of various forms of energy use. The author covers energy generation by solar, nuclear, hydroelectric, wind, tidal, and geothermal means, and clearly explains concepts related to each alternative, suggesting technical potential as well as concerns. Chapters 1 to 3 provide an overview of energy fundamentals and traditional sources of energy generation. Chapters 4, 5, and 16 examine nuclear power in terms of its costs and benefits. Chapters 6 to 9 explore several energy alternatives, such as solar and wind, their means of storage, and energy conservation. Chapters 10 to 12 examine the human aspects of energy generation and use, from transportation to plant and food production. Chapters 13 to 15 look at the negative effects of current patterns of energy use including water, air, and noise pollution. The final chapter examines the threats of nuclear war and gives examples of the improper use of nuclear energy.

33. Leach, Gerald, Lorenz Jarass, Gustav Obermair, and Lutz Hoffman. *Energy and Growth: A Comparison of 13 Industrial and Developing Countries*. London: Butterworths, 1986.

Keywords: case studies, consumption, economics, fuels, planning, supply and demand

The authors of this volume maintain that sound energy planning must be based on an accurate understanding of energy demand and the social and economic interactions that organize societies.

With this approach in mind, they compare five industrialized, four middle-income, and four low-income countries in terms of their patterns of energy use and changes experienced since the oil crisis of the 1970s. They compare the energy sector, of each country in terms of energy use in agriculture, transport and housing; energy prices; and in terms of more conventional aggregate measures of energy production, trade, and conversions. Chapter 1 is entitled "Energy and Economic Structures" and examines the energy crisis, the role of biomass energy, economic structures, levels of development, and models of the energy system. Chapter 2, "Aggregate Country Comparisons," examines energy-population ratios and energy-GDP intensities. Chapter 3, "Energy Supply and Trade," discusses energy self-sufficiency, domestic energy production, energy trade and its economic impacts, energy transformation and distribution, and the electricity supply sector. Chapter 4, "Disaggregation," presents an "ideal" model for disaggregation, as well as examples of disaggregation by energy intensity and GDP share. Chapter 5, "Fuel Shares," examines fuel shares of primary and final energy use in both the industrial and residential sector. Chapter 6, "Energy Prices," analyzes international and domestic crude oil prices. Chapter 7, "Industry," deals with the changing share of industrial GDP, industry-specific energy intensities, and the iron, steel, cement, and aluminum sectors. Chapter 8, "Agriculture," discusses energy shares in agriculture, conditions of energy applications, and energy intensity and costs in agriculture. Chapter 9, "Transport," analyzes data problems, transport for production and consumption, and reduction of transport energy intensities. Chapter 10, "Households," discusses household surveys and data problems, energy shares, per capita energy use and space heating, energy-income intensities, survey comparisons, effects of household size, budget shares, energy conservation, and a transition from biomass fuels. Chapter 11, "Summary and Conclusions," discusses energy consumption of different economies, sectorial analysis, the impact of energy prices, and supply and demand factors.

34. Leach, Gerald. *Household Energy in South Asia*. London: Elsevier Applied Science, 1987.

Keywords: case studies, consumption, economics, energy analysis, policy, poverty alleviation, supply and demand

The main goal of this book is to enhance the reader's understanding of household energy and its linkages. The author analyzes household energy use, supply, prices, and other relevant factors within the context of South Asia. Chapter 1 illustrates that the fuelwood problem is another manifestation of poverty in rural areas, and as such it must be addressed in a broad policy context that pays close attention to the characteristics of each specific location and social group. Chapter 2 demonstrates that income, household size, and settlement size influence the amount and type of energy used at the individual household level. Chapter 3 analyzes biofuel consumption patterns, as well as price fluctuations of modern household fuels and urban firewood prices. Chapter 4 uses empirical data from Indian villages to explore the reasons behind the marked variations in fuel consumption patterns. Chapter 5 examines the massive transition from biofuels to modern fuels observed throughout urban India during the 1980s. The concluding chapter discusses some of the most salient policy issues related to household energy.

35. Lin, Xiannuan. *China's Energy Strategy: Economic Structure, Technological Choices, and Energy Consumption.* Westport, Conn: Praeger, 1996.

Keywords: case studies, conservation, policy

The research project entitled "Factors Behind the Fall in China's Energy Intensity" forms the basis of this book. Its purpose was to explain the drop in China's energy intensity that occurred in the 1980s and to further determine major sources of energy savings. The energy-use changes in China's economy from 1981 to 1987 are analyzed within the general framework of input-output economics. Energy-use changes within a unified and clearly specified framework of structural decomposition analysis are explored, with a consistent set of concepts, definitions, assumptions, and computations. The analysis covers all sectors of the economy, including the material-production sectors and the service-production sectors. Additionally, the intersectoral input-output linkages are traced and account for both direct and indirect energy use, which provides an understanding not only of the direct results of an energy conservation measure, but also its effects on intersectoral linkages. Finally, the book places structural decomposition analysis into the broad context of changing economic and institutional conditions in China and links energy-use changes with the economic reform program and changes in government policies in the 1980s. In this analysis of energy reduction strategies in China and their strengths and shortcomings, suggestions for further reduction strategies are also given with a case study in chapter 6 of the iron and steel industry.

36. Lovins, Amory B. *Soft Energy Paths: Toward a Durable Peace.* San Francisco: Friends of the Earth International, 1977.

Keywords: alternative energy, economics, nuclear

This classic work explores the fallacy behind the traditional method of determining energy needs by means of GNP/GDP analysis. The author deals with issues such as the desirable rate of economic growth and the view that there is a growing need for increased energy use. Throughout the book, the author promotes the view that people are more important than goods. This view translates into a different understanding of energy, technology, and economic activity as means as opposed to ends; and that, as such, these concepts alone are not accurate measures of social welfare. The book is divided into three parts. Part 1, "Concepts," gives an overview of the author's thesis which examines the old adage "technology is the answer!" to which the author adds "what was the question?" Part 2, "Numbers," examines "soft" versus "hard" technologies and the costs of each—in both long and short terms. This part also provides clear definitions and calculations. Soft energy paths are characterized by the fact that they are based on renewable resources, are diverse, flexible, transparent to the public, of appropriate scale, and use appropriate energy quality—matching the minimum grade of energy necessary to perform the level of work desired. Such paths are, moreover, much more compatible with democratic values, and contribute to world peace by cutting reliance on foreign oil. Part 3, "Toward a Durable Peace," looks into the future by questioning the validity of nuclear power and examining the sociopolitical requirements of following a soft path.

37. Lovins, Amory B. *World Energy Strategies: Facts, Issues, and Options*. New York: Harper Colophon Books, 1980.

Keywords: energy analysis, fuels

This book avoids easy answers and concentrates on a careful technical assessment regarding the nature and magnitude of constraints of current energy resources. Within this context, the author identifies long-term energy options and short-term actions that must be avoided if we are to preserve these options. He notes that the rapid energy growth rates that most industrial countries have long maintained cannot continue for much longer. Therefore, industrialized nations should immediately undertake lasting and fundamental measures. The author believes that incentives should be implemented so a diverse range of "unconventional" energy technologies that are "fission-free" can develop. The book emphasizes the implementation of changes in lifestyle, as well as changes that are not technical or economic but are rather social and ethical, in all policy decisions. The final part examines world energy conversion, fossil and non-fossil fuels, energy income, ethics, and risk.

38. Lovins, Amory B., and L. Hunter Lovins. "Least-Cost Climatic Stabilization." *Annual Review of Energy and the Environment* 16 (1991): 433-531.

Keywords: environmental, economics, efficiency, case studies

Human-induced climate change is a direct consequence of the inefficient use of energy and resources. This article advocates the use of advanced technologies to stabilize the global climate and support global economic activities. Several innovative resource-saving techniques that are cost-effective and generate fewer greenhouse emissions are identified. These may be implemented in areas such as energy, forestry, and agriculture. The article provides a summary of new forms of analysis that conclude that measures to deal with global warming can be implemented in many regions at a negative net cost. However, the authors note that the specific details of these measures are directly related to the culture and geography of each particular region. The article identifies and analyzes measures to use resources more efficiently in OECD countries, developing nations, and in the Russian Federation of Independent States. It specifically analyzes issues of trade, technological transfers, and the development of effective models in order to develop an implementation agenda.

39. Mabey, Nick, Stephen Hall, Clare Smith, and Sujuta Gupta. *Argument in the Greenhouse: International Economics of Controlling Global Warming*. London: Routledge, 1997.

Keywords: case studies, developing countries, economics, efficiency, environment, policy

Unlike many other technological assessments of global warming, this book attempts to give a descriptive analysis of the consequences of different ways to control climate change. The authors present the economic, political, and legal framework of the FCCC (Framework Convention on Climate Change) as encompassing the quantitative results of the analysis. Quantitative, econometrically based techniques are used to model the cost to developed countries of complying with future commitments under the FCCC and how the distribution of costs will affect the potential for agreement in the future. Following the introduction in Chapter 1, a qualitative analysis of some of the main policy issues is presented, as well as the reasons behind the need to quantify different effects. Part 2, which includes chapters 3 to 6, looks at the economic technicalities of constructing an EGEM (Environmental Global Econometric Model), which is the model used in this study. Topics include modeling issues, past work, empirical modeling of energy demand responses, modeling the macroeconomic effects of carbon abatement, and a study of India in terms of carbon abatement in developing countries. Part 3 addresses the international economics of climate change. Chapter 7 covers the theoretical question of optimizing carbon dioxide abatement; chapter 8 models how uncertainty, learning, and strategic behavior interact; chapter 9 questions the consequences of the FCCC's commitment to unilateral control of carbon emissions in the OECD; chapter 10 looks at the domestic political economy of limiting carbon emissions; and finally, chapter 11 covers the means by which different policy instruments can contribute to coordinating an efficient and stable carbon abatement treaty between the major OECD countries. Part 4 serves as an overview, followed by a summary and conclusions.

40. Majumdar, Shyamal K., ed. *Environmental Consequences of Energy Production: Problems and Prospects*. Easton, Pa.: Pennsylvania Academy of Science, 1987.

Keywords: economics, electricity, fuels, generation, hydroelectric, nuclear, planning

Several international experts contributed papers to this book, which analyzes the environmental consequences related to the cycle of energy production and use. It explores the implications of implementing alternative sources of energy generation at the global, regional, and national levels. The goal of the authors is not to merely describe environmental problems but to provide practical solutions to them. This collection of papers written by scientists, engineers, and social scientists is divided into five sections. Section 1 examines the environmental problems related to the production of coal, oil, natural gas, and uranium. Section 2 covers the environmental effects of energy production from fossil fuels. Section 3 discusses the environmental impacts of hydroelectric power. Section 4 examines the environmental consequences and management challenges of nuclear power generation. Finally, section 5 looks at environmental legislation, as well as economic and social considerations.

41. Meyers, Stephen, and Lee Schipper. "World Energy Use in the 1970s and 1980s: Exploring the Changes." *Annual Review of Energy and the Environment* 17 (1992): 463-505.

Keywords: consumption, environmental

The authors note that the dominant patterns of supply and use of energy create severe problems and policy challenges for individual nations and the global community. Problems such as climate change, international conflicts over oil supplies and nuclear proliferation, and the intolerable economic and social costs of satisfying basic human needs are all related to the predominant global patterns of energy use. This article summarizes an analysis of trends in energy use by sector for most of the world's major energy consuming countries. The period from 1970 to 1988 was analyzed by a group at the Lawrence Berkeley Laboratory. The purpose of this analysis is to gain an understanding of the reasons behind changes in energy use. This in turn should enhance our ability to predict future problems and increase the effectiveness of potential solutions.

42. Miller, E. Willard, and Ruby M. Miller. *Energy and American Society*. Santa Barbara, Calif.: ABC/CLIO, 1993.

Keywords: case studies, economics, policy

The rich energy resources of the United States have been responsible for its strong industrial position, high standard of living, and economic strength. This guide examines the history, consumption, and availability of energy resources in America and the influence that energy has on the development of modern-day society. It looks at the current state of U.S. energy resources, related government regulations, the reasons for the slow progress in developing alternative resources, and the economic and social problems that could arise from an energy crisis. This book provides both an in-depth exploration of the above topics and a one-step information source complete with facts of all sorts.

43. Morris, David. *Self-Reliant Cities: Energy and the Transformation of Urban America*. San Francisco: Sierra Club Books, 1982.

Keywords: cities, planning, sustainability

American cities have been transformed by changing sources and forms of energy. Part 1, "Losing Control," chronicles the development of small, independent, self-sufficient villages into large cities dependent on imported fuels and materials. This part focuses on the period from 1870 to 1970 during which oil prices fell continuously. The cities' increasing dependence on remote corporations and governments and the loss of citizens' economic and political power are emphasized. Part 2, "Gaining Autonomy," reports on cities' attempts at gaining self-reliance in the "age of expensive energy." Technological, institutional, and financial changes that are leading to increased local control and sustainability are considered. Morris envisages cities with humanly scaled energy systems and completely restructured waste disposal and transportation systems financed by energy utilities and municipal corporations.

44. Morrison, Denton E., and Dora G. Lodwick. "The Social Impacts of Soft and Hard Energy Systems: The Lovins' Claims as a Social Science Challenge." *Annual Review of Energy* 6 (1981): 357-78.

Keywords: economics, energy analysis, policy

The authors note that Amory Lovins has made an extremely important and immensely influential contribution to the energy debate. The key feature of his contribution has been to pose the question of energy policy in terms of "soft" and "hard" energy paths. The first part of this article analyzes the characteristics, implications, and social consequences of the soft energy path enunciated by Amory Lovins. The second part of the article presents a set of practical methodological considerations that can be used to design research strategies to study and test different aspects of the soft energy path concept.

45. Munasinghe, Mohan. *Energy Analysis and Policy*. London: Butterworths, 1990.

Keywords: case studies, energy analysis, conservation, economics, efficiency, developing countries, fuels, planning, policy, poverty alleviation, supply and demand

In this book, Munasinghe analyzes the ability of energy sector organizations in developing countries to achieve economic efficiency in their investment decisions and operations through the practice of rational economic principles. He examines the integration of supply capacity enhancement with a demand-management approach, in order to assess the efficient use of resources for development. He concludes by discussing the development of the energy sector in terms of maximizing the welfare of society as it grows, particularly with regard to the growing gap that exists between rich and poor in developing nations. In addition to dealing with these issues, the book provides a framework for analysts and policy-makers to pursue integrated energy planning while keeping in mind the overall objectives of growth and equity.

46. Munasinghe, Mohan, and Peter Meier. *Energy Policy Analysis and Modelling*. London: Cambridge University Press, 1993.

Keywords: case studies, electricity, energy analysis, environmental, planning, policy

The authors observe that the fact that energy is directly related to social and economic activities makes energy planning and policy analysis an essential area of study. This volume explains strategies used by energy planners, such as the concept of INEP (Integrated National Energy Planning) and spreadsheet, linear planning, and optimization models. The volume introduces a modeling scheme to facilitate energy planning and policy analysis in developing nations. This conceptual framework, together with environmental considerations, is used to analyze the electricity and fuelwood sub-sectors and is provided as background information for a discussion on integration problems and policy implementation issues. The arguments presented are supported by case studies from Egypt, West Africa, Sudan, Pakistan, Colombia, India, Sri Lanka, and Morocco.

47. Myers, V. Norman, ed. *Gaia: An Atlas of Planet Management*. New York: Anchor Books, 1993.

Keywords: alternative energy, environment, renewable

The present global crisis is mapped out and analyzed in this account of a living planet at a critical point in its evolution. This comprehensive study looks at seven topics: land, ocean, elements, evolution, humankind, civilization, and management. Within these sections, areas of concern are investigated and possible solutions proposed. A large part of this study is concerned with energy and its sources. In terms of land, the amount of energy used by nations around the world for agriculture is looked at. A brief look is taken at ocean-generated energy, such as wave energy, ocean thermal energy conversion, and spiral wave power. The section covering "elemental potential" deals with the energy issue in more depth. Topics covered include the human energy budget, biomass, tidal energy, solar energy, geothermal energy, hydropower and power from the sea, wind, nuclear, oil, coal, and natural gas. Also discussed are the world distribution of energy resources, world mineral reserves, the oil crisis, the fuelwood crisis, the nuclear dilemma, the new energy path, an energy-efficient future, and managing energy in the south.

48. Newson, Malcolm. *Managing the Human Impact on the Natural Environment: Patterns and Processes*. London: Belhaven Press, 1992.

Keywords: environmental, nuclear, supply and demand

Air, land, and water pollution are the focus of this book, as well as the source, pathway, and target of pollutants, and the spatial and temporal processes which help nature deal with pollution: dilution, dispersion, and decay. Part 3, entitled "Futures," deals with energy issues. Chapter 11 looks at radiation and the environment, touching on nuclear power and related concerns. Chapter 12 examines the global environmental implications for future energy supply and use in terms of the history and geography of energy development and energy-related pollution; energy reserves and use; energy demands and development; energy supply; world trade in energy; fuel technology; pollution related to energy generation; and future energy supply patterns.

49. Organization for Economic Co-operation and Development. *Environmental Policies for Cities in the 1990s*. Paris: OECD, 1990.

Keywords: case studies, CHP, cities, conservation, economics, environmental, policy, transportation

A conference that took place in Berlin, Germany, in 1989 served as the foundation for this volume, which probes several existing urban environmental improvement policies. It looks at ways to improve the coordination of policies that have an environmental impact on cities, and presents policy instruments that can be used by national, regional, and local governments. It also looks at recent progress in urban rehabilitation, urban transport, and urban energy management, as well as means of improving these three areas. Chapter 1 concentrates on the nature of the urban environmental challenge. Chapter 2 focuses on a policy framework for the urban environment in terms of necessity, aims, and initiatives and the future of sustainable development for cities. In chapter 3, organizational integration and economics as keys to success are analyzed. Finally, chapter 4 addresses three issues: urban energy, urban transport, and urban rehabilitation, with an emphasis on the identification of innovative approaches and what can be learned from financing, short- and long-term impacts, and political feasibility. There are two relevant insets in chapter 4. The first, number ten, deals with the role of combined heat and power (CHP) in energy conservation and emission reduction in Denmark, and the second with least-cost utility planning in Seattle, Washington.

50. Organization for Economic Co-operation and Development. *Urban Energy Handbook: Good Local Practice.* Paris: OECD, 1995.

Keywords: environmental, cities, policy, sustainability

This book focuses on environmental improvement through urban energy management at the local level and the relation of action at this level to larger-scale policy concerns. Urban issues such as policy integration, cogeneration and district heating, renewable forms of energy generation, education and training, transport, and energy indicators are all discussed in relation to the needs and responses of citizens, city officials, and policy-makers. Chapter 1, "Sustainable Energy in Cities: Strategic Approaches," examines the problem of global warming. This chapter emphasizes the importance of local governments in combating this problem, as well as in improving the competitive position of cities by attracting investment, businesses, and high-skilled workers. Numerous case studies represent a substantial part of the book's content. These case studies, drawn from cities around the world, discuss technical initiatives, policy options, results of implementation of solutions, and—for those cities still in the evaluation phase—preliminary indications of progress. They provide excellent practical examples of policy implementation, which are complemented by numerous contacts and sources for further investigation.

51. Owens, Susan. *Energy, Planning, and Urban Form.* London: Pion, 1986.

Keywords: case studies, CHP, conservation, efficiency, energy efficient buildings, planning, policy, transportation

This book attempts to bridge the gap between theory and practice in the energy field. It treats energy not as an exclusive determining factor in urban change, but as an important force behind change. Emphasis is placed on barriers to formulating and implementing relevant policies; real problems from the point of view of those in planning practice; and more theoretical aspects of identifying energy-efficient spatial structures. Spatial responses to energy constraints are looked at with regards to urban models and the relation between residential energy consumption and spatial structure. Reducing transport energy requirements is studied with regard to energy-efficient structure, on various regional scales. The reduction of energy requirements is also looked at in terms of buildings: energy efficiency and built form, renewable energy and spatial structure, and combined heat and power generation. In a final synthesis, energy-efficient environments are examined in terms of efficient structures, the potential for energy conservation and savings, and the policy implications and constraints of energy and planning. Finally, energy-integrated planning in practice is analyzed in several case studies: Davis, California, and Portland, Oregon, in the United States; London and Milton Keynes in the United Kingdom; Melbourne, Australia; and Denmark.

52. Patterson, Walt. *Rebuilding Romania: Energy, Efficiency, and the Economic Transition*. London: Earthscan, 1994.

Keywords: case studies, economics, efficiency, policy

Romania is a country seldom mentioned in discussions about energy advances and new developments in the field; in fact, it enters the conversation more frequently in terms of countries lacking in energy use and maximization of energy potential. Things are changing, however, as Romania pushes to rebuild and to make a place for itself in the international setting. This work begins with a brief but comprehensive look at Romania's political and energy history and the events that led to its decline, ending with the execution of Ceausescu. It also presents a broader introduction to Romania which prepares the reader for what is to come. Following this background information, chapter 2 describes the rebuilding process in terms of the present and projected structure of Romanian energy supply and use. Romania's energy framework is addressed in chapter 3 with a look at its key energy institutions and their activities. Financing transition is the subject for chapter 4, while chapters 5 and 6 focus on promoting and investing for energy efficiency. Chapter 7 identifies opportunities for improvement with a look at international cooperation, and chapter 8 discusses energy efficiency and policy.

53. Pimentel, David, and Carl W. Hall. *Food and Energy Resources*. New York: Academic Press, 1984.

Keywords: alternative energy, developing countries, economics, environmental, policy, renewable

The growth rate and the current size of the human population constitute an ongoing, and growing, challenge for the supply of food and energy resources. This challenge is further complicated by issues of sustainability and the finite nature of fossil resources. The goal of this volume is to provide, for a wide range of specialists, a clear understanding of the relationships between food and energy issues, and to provide background information for the design of strategies to deal effectively with these issues. The book is divided into ten chapters, written by an interdisciplinary group of authors, and covers the following topics: energy flow in the food system; energy sources and conversions relating to food; the role of energy in world agriculture and food availability; food for people; energy use in crop systems in Northwestern China; energy and food relationships in developing countries: a perspective from the social sciences, ethics, economics, energy, and food conversion systems; solar energy applications in agriculture; biomass energy, and food conflicts; alcohol from corn grain; and the relation of residues to prices, land use, and conservation.

54. Rudolph, Richard, and Scott Ridley. *Power Struggle: The Hundred Year War Over Electricity*. New York: Harper & Row, 1986.

Keywords: economics

The authors explore the intrigues related to the century-long struggle between public and private interests to dominate the supply of energy in the United States. They note that this struggle shapes the political landscape, industrialization patterns, environmental quality, and the concentration of economic power in the United States. Their analysis focuses on the history of this energy struggle and its cyclical influence on the future development of alternative energy paths. The authors believe that the outcome of this struggle depends mostly on political forces rather than technological developments.

55. Samuels, Robert, and Deo K. Prasad, eds. *Global Warming and the Built Environment*. London: E & FN Spon, 1994.

Keywords: environmental, cities, efficiency, energy efficient buildings, sustainability, transportation

The question is no longer whether we contribute to environmental deterioration through our built environment, but to what degree. This book attempts to make us aware of the relationship between the two, as well as between the many elements that make up the built environment, in working towards sustainable development. It looks at the contributions we can make through the use of renewable energy, the potential for energy efficiency in terms of commercial buildings, changes in architectural form and design, the idea of "efficiency with sufficiency," and the efficiency and costs of various building materials. Chapter 1 calls for a move towards the "common good," rather than the "individual good" on the path to sustainable development. Chapter 2 looks at environmental auditing and the importance of corporate involvement in environmental issues. In chapter 3 there is a discussion of the need for honesty in terms of the capabilities, benefits and impacts of renewable energies in working towards sustainable development. Chapter 4 focuses on issues surrounding the greenhouse effect, especially its impact on future generations. This is followed by a discussion of the contributions that urban design and transportation systems can make to sustainable development. A survey of thirty-one cities allows for many suggestions for the modification of urban design, and some means of action are discussed in chapter 5. Chapter 6 suggests an alternative to the ideas set forth in chapter 5, in terms of creating a "sustainable suburbia instead of modifying urban design." Chapter 7 deals with the complex relationship between energy and architectural form in working towards improving living conditions in the built environment as well as increasing energy efficiency. Chapter 8 stresses the important role that architects must adopt in creating proactive designs and looks at several examples of such architecture. Chapter 9 examines energy efficiency in non-residential buildings. Chapter 10 deals with energy efficiency in residential buildings and how developed countries still have a long way to go with improving efficiency. Chapter 11 discusses technological options and sustainable energy welfare, and chapter 12, materials selection and energy efficiency, as well as estimates of energy's capital cost in buildings.

56. Schipper, Lee. *Energy Efficiency and Human Activity: Past Trends, Future Prospects*. New York: Cambridge University Press, 1992.

Keywords: efficiency, consumption, policy, sustainability, transportation

This is a detailed study of world energy use over the past twenty years and future prospects. Means of maintaining growth in consumption are discussed, as they pertain to a "need" to meet environmental and economic development goals. The authors and their colleagues at Lawrence Berkeley Laboratory worked together in order to present this study on energy use and the forces shaping it in the industrial, developing, and formerly planned economies. An overview of the potential for improving energy efficiency is put forward and the policies that could help with such a realization are addressed. Suggestions made are: "strong action by governments and the private sector," and "considering the full range of factors that will shape realization of the energy efficiency potential around the world." The first of the three sections looks at past trends in terms of world energy since 1970 and historic trends in manufacturing, transportation, the residential sector, and the service sector. The second examines future prospects, including the outlook for activity and structural change related to energy intensities and scenarios of future energy intensities. In the third section, ensuring a sustainable future is the key concern in terms of encouraging energy efficiency, thorough policies and programs, and altering the relationship between energy and human activity.

57. Steen, Nicola, ed. *Sustainable Development and the Energy Industries: Implementation and Impacts of Environmental Legislation*. London: Earthscan, 1994.

Keywords: efficiency, policy, sustainability

The workshop held by the Energy and Environmental Programme to look at environmental legislation and energy industries served as the foundation for this book. It is divided into four parts: "The Challenge of Sustainable Development," "The Context and Constraints," "Energy Industry Experience and Perspectives," and "Opportunities and Strategies." Part 1 contains an introductory address that attempts to clarify the concepts associated with sustainable development as well as the dilemmas raised in implementing it. Part 2 includes an article by Lars J. Nilsson and Thomas B. Johansson which maintains that end-of-pipe solutions will not be adequate to meet present environmental goals, and that a better option is energy efficiency improvements and use of renewable sources of fuel and electricity. Part 4 contains an article by von Weizsacker claim that, although we must reduce global greenhouse gas emissions, nuclear and renewable energy will not be able to provide enough power for our needs, and the only reasonable solution is to increase energy productivity. The pros and cons of trying to achieve more efficient energy intensities through bureaucratic measures versus the free market are then explored in detail. The choice of one or the other, or a mixture of the two, will be determined by existing national conditions.

58. Stern, Paul C., and Elliot Aronson. *Energy Use: The Human Dimension*. New York: W.H. Freeman & Co., 1984.

Keywords: case studies, consumption, policy

The members of the Committee on the Behavioral and Social Aspects of Energy Consumption and Production worked together in the production of this book. Their goal was to review literature in the behavioral and social sciences that was relevant to an understanding of energy production and consumption in the United States. Energy policy issues became increasingly important as they drew on work from these fields. Due to the large scope of the material in this area of study, not everything could be closely examined. It was decided that it would be more beneficial to treat selected contributions in depth rather than looking at a larger number of works in a more superficial way. It was hoped that the results from several social and behavioral sciences would increase the understanding of policy-makers and citizens about ways people relate to the energy system and how different factors affect this whole. Stern and Aronson investigate topics such as the human dimension: our great dependency on energy systems; some barriers to energy efficiency; individuals and households as energy users; and organizations and energy consumption.

59. Tatum, Jesse S. *Energy Possibilities: Rethinking Alternatives and the Choice-Making Process*. Albany: State University of New York Press, 1995.

Keywords: economics, energy analysis

The belief that "present energy practices and policies in the Western World do not reflect the best interests of ordinary people" is Tatum's central thesis. Traditional methods of dealing with the energy problem are described as inadequate and inappropriate. Tatum believes that energy choices are inextricably linked to how people decide to live in the world. The book moves from theoretical considerations to a discussion of unique practical responses to energy concerns with the intention of empowering the reader. Chapter 1 begins by raising broad questions about the validity and wisdom of traditional energy policies and policy methods. Chapter 2 looks at energy technologies and their profound influence on our way of life, as well as the shortcomings of the "technological fix" solutions we have traditionally relied on. Chapter 3 looks at responses from the engineering and economic points of view and again challenges traditional methods as being hopelessly blinkered. Chapter 4 discusses the differing views of experts, the appropriateness of various economic measures and objectives, other disciplinary perspectives, and the radical impoverishment of imagination in conventional thought regarding energy. Chapter 5, "The Shaping of Responses," looks at alternative responses, the "promise of technology," and issues of empowerment and responsibility. The "home power" movement, how it began, the technology associated with it, and new developments are discussed in chapter 6. Chapter 7 concludes with an analysis of the various policy options open to us and the social arrangements that are associated with each of these options.

60. Toke, David. *The Low Cost Planet*. London: Pluto Press, 1995.

Keywords: economics, efficiency, environmental, fuels, nuclear, pollution prevention, renewable

A common belief is that repairing environmental damage will raise the costs of supplying energy to consumers, thus increasing costs overall. This book takes a close look at the accuracy of this line of thinking. Several issues are investigated, such as energy efficiency, fossil fuels, nuclear power, pollution problems, and renewable energy. The author discusses some solutions and argues that even the largest energy and environmental problems can be dealt with while avoiding additional costs to the consumer. Chapters 2 and 3 look at both the pollution problem and the resource problem. Some main solutions for the energy debate are covered in chapter 4, while chapter 5 discusses the establishment of a common standard for assessing the costs of different energy options. Chapter 6 covers the use of natural gas as a possible way of addressing environmental and resource problems. The technology involved in the implementation of energy efficiency and the policy strategies needed to make it come about are the topics of chapter 7. Chapter 8 covers cars and ways to fight the pollution they cause. Chapter 9 looks at coal use and ways of lessening its environmental impact. Chapter 10 discusses methods of removing from the air the carbon dioxide emitted by fossil fuel burning. Chapters 11 and 12 look at nuclear power, and renewable energy, and their potential, costs, and environmental impacts. The final chapter reviews various policy options in terms of their economic and technological feasibility.

61. Vanderburg, Willem H. *The Labyrinth of Technology*. Toronto: University of Toronto Press, 2000.

Keywords: sustainability, energy, urban metabolism, technology

A detailed diagnostic study of engineering education in particular and professional education in general, along with a comparison of conventional and state-of-the-art practices for dealing with the social and environmental implications of technology, reveals a professional ethos that concentrates on technological development in terms of performance ratios such as efficiency, productivity, profitability, and GDP that masquerade as social values. Such input-output ratios provide us with no indication as to whether any improvement in performance is partly or wholly achieved at the expense of human life, society, and the biosphere. These considerations are attended to in an end-of-pipe or after-the-fact manner resulting from the intellectual division of labor and corresponding institutions that first create problems and then seek to resolve them. These problems are compounded by an economic bookkeeping that makes no distinction between gross and net wealth creation.

The above diagnosis leads to a possible prescription. Preventive approaches gather information about how technology interacts with and depends on human life, society, and the biosphere in order to adjust design and decision-making to ensure that our goals and aspirations are not compromised. After developing a conceptual framework for preventive approaches (including individual and organizational prerequisites), basic intellectual tools and values are set out. Next, three dimensions of sustainability form the basis for decision matrices to ensure that solving a problem in one area will not create others elsewhere. Finally, preventive approaches are developed in four areas of application: materials and production, energy, work, and cities.

62. Wallace, David. *Sustainable Industrialization*. London: The Royal Institute of International Affairs, 1996.

Keywords: case studies, conservation, economics, environmental, fuels, sustainability

The struggle for sustainable development is becoming increasingly difficult because of worldwide economic liberalization and subsequent rapid industrialization. Wallace examines the role played by energy in the second and third chapters of this book. Chapter 2 looks at national strategies for sustainable development in the Netherlands, Germany, and Denmark as well as corporate efforts towards the same goal. In addressing local and regional environmental problems, the Dutch National Environmental Policy Plan (NEPP) regards the conservation of energy and the use of cleaner energy sources as one of three ways to achieve their goals. Germany's agenda for working towards sustainable development, although not quite as comprehensive as the Dutch system, is closing consumption cycles in terms of materials and energy through recycling. Chapter 3 deals with energy specifically within the framework of the West's record on sustainable development, among other things. Some of the topics looked at are historical trends in energy intensity in some countries; estimated energy efficiencies in OECD countries; theoretical energy efficiencies for major fuels; and substitution of steel technologies in the United States.

63. Woodward, Alison E., Jerry Ellig, and Tom R. Burns. *Municipal Entrepreneurship and Energy Policy: A Five Nation Study of Politics, Innovation and Social Change.* Langhorne: Gordon and Breach Science Publishers, 1994.

Keywords: case studies, conservation, policy

Innovation has many political and social repercussions. The organizational developments and social conflicts that result are examined in this book. In making such an analysis, all factors have to be considered, particularly the "interplay between actors, events and their organizational, cultural and material contexts." The work is divided into three parts. Part 1 serves as an introduction, to "action, entrepreneurship and energy" and "research concepts and design." Part 2 contains the case studies of six different regions from around the world: Nysted, Denmark; Metz, France; Saarbrucken, Germany; Goteborg, Sweden; and Davis, California, USA. Questions asked in the study were related to innovation for energy saving, entrepreneurs, motivation, institutional settings, policies and programs, and factors that facilitated or constrained innovation. Part 3 is an analysis of municipal entrepreneurship, where energy developments are compared and generalizations drawn from this comparison.

64. World Energy Council. *Energy for Tomorrow's World: The Realities, the Real Options and the Agenda for Achievement.* New York: St. Martin's Press, 1993.

Keywords: case studies, developing countries, economics, efficiency, environmental, supply and demand

This book was the result of a project that focused on important contemporary energy issues within various regions of the world in order to aid economic, technical, environmental, social, and institutional development. It also used information and issues discussed at the fifteenth WEC Congress of 1992, and addresses important issues that will affect the way that energy is generated, supplied, and used in the future. These issues include population growth, economic and social development, access to sufficient energy for the developing world, local and regional environmental impacts, global climate change, efficiency of energy supply and use, financial and institutional issues, and technological innovation and dissemination. Part 1 looks at patterns of energy use in terms of policies, population, economic growth, demand, pricing, competition, and technology. Energy supply is covered in terms of fossil fuel resources, non-fossil energy supplies, constraints on supply, the financing of future energy supplies, and the price of energy.

Energy efficiency and conservation are addressed, as well as energy and the environment where the effects of environmental regulation on various forms of energy are taken into account. Part 2 discusses the priorities of various regions: North America, Latin America, the Caribbean, Western Europe, Central and Eastern Europe, the Commonwealth of Independent States, Georgia and the Baltic States, the Middle East, North Africa, Sub-Saharan Africa, South Asia, and the Pacific. Finally, major concerns, such as energy poverty, resource and geopolitical constraints on energy supply and demand, pollution and degradation of the environment, institutional issues, and conclusions and recommendations are discussed in part 3.

Restructuring the Network for Greater Sustainability

Energy Supply

Non-Renewable Energy Sources

   This section contains works that deal preventively with energy supply through conventional resources. Again, there are a few works that are also relevant for subsequent sections. The first entry (No. 64), for instance, makes contributions to an analysis of renewables as well as for policy. In some cases the editors had to make a difficult judgement call in terms of where to include a particular work. In such cases the several thrusts of the work were assessed for their prominence, and the work was placed in the category that was deemed to be relevant to its most important theme. For instance, Paul Rosenberg's (No. 76) *Alternative Energy Handbook* makes significant contributions to the following subsection on renewables. Its discussion of nonrenewable supply strategies, however, outweighed this other contribution.

65. Babus'Haq, R. F. and S. D. Probert. "Combined Heat-And-Power Implementation in the UK: Past, Present, and Prospective Developments." *Applied Energy* 53 (1996): 47-76.

Keywords: case studies, CHP, fuels, renewable

   This article argues that "the world is insufficiently concerned with maintaining its long-term energy supplies." Because it is thought that a severe reduction in the availability of cheap fossil fuels is likely to occur in the mid-twenty-first century, several steps should be taken in the UK to ensure the sustainability of adequate power supplies. A variety of fuels are considered including oil, coal, natural gas, refuse, sewage, nuclear power, and renewables. The article also looks at the pace of adoption of CHP in the UK, which, despite being "thermodynamically attractive," has been disappointingly slow. A historical record of the evolution of CHP as well as district heating in the UK is presented. Recent developments and a possible future scenario for energy supplies are also outlined.

66. Brown, Kevin. "History of CHP Developments and Current Trends." *Applied Energy* 53 (1996): 11-22.

Keywords: case studies, CHP, efficiency

This paper examines the perspective of the British Government on the desirable role of combined heat and power systems (CHP) in relation to the country's energy supplies. It looks at government initiatives aimed at encouraging greater market penetration for CHP and the potential for future growth of the national CHP industry. The paper is divided into the following sections: Energy Efficiency Office (EEO), industrial CHP, building CHP, biofuel/waste CHP, best practice program, current technology and markets, other players in the market, and the future of CHP.

67. Flavin, Christopher, and Nicholas Lenssen. "Reshaping the Electrical Power Industry." *Energy Policy* 22, 12 (1994): 1029-44.

Keywords: general, electricity, supply and demand

This article begins with a brief history of the electricity industry, from the first electricity sold by Thomas Edison to the present, a period characterized by rapid transformations in the global economy as well as in the environment. The authors maintain that most contemporary vertically integrated utilities will be restructured or divided during the next decade. These changes will result in the creation of a commodity market in power generation and transmission and a "competitive services market at the local distribution level." They believe that these transformations of the power industry are caused by larger privatization and deregulation trends at the macro level and by technological advances in energy production and consumption. The article concludes that "devices such as fuel cells, photovoltaics, and flywheels will open the way to a more decentralized power industry in which electricity generation and storage facilities are increasingly located in the customers' own facilities—integrated and controlled by new digital communication systems."

68. Hill, Robert. "Environmental Implications." *Applied Energy* 53 (1996): 89-117.

Keywords: efficiency, CHP, environmental

Robert Hill explores the consequences of replacing fossil-fuelled central electricity generating systems with combined heat and power (CHP) systems. The article begins by developing a methodological framework to examine the external costs of energy technologies. The effects on the environment and other problems are assessed by means of a fuel-cycle analysis. The following generation technologies are described: coal-central stations; combined-cycle gas turbines; heating boilers; small scale CHP; large scale CHP; gas turbine CHP units; reciprocating engines; and fuel cells. The conversion efficiencies of all these technologies are assessed. The article argues that one of the greatest advantages of CHP is its efficiency in terms of primary energy resource utilization, and provides a comparison of gas emissions from central stations and CHP plants.

69. International Energy Agency. *Biofuels*. Paris: OECD/IEA, 1994.

Keywords: general, economics, environment, fuels

Biofuels have been gaining interest as alternative fuels, gasoline extenders, or additives and as a means of generating electricity or heat. This text focuses solely on biofuels in their various forms and origins, their uses, potential, benefits and, in some cases, barriers to more widespread use, as well as their impact in terms of greenhouse gas abatement and cost benefits. Section 2 addresses the motivation for using biofuels and their use in response to various policy objectives, specifically in terms of energy and environment, as well as agriculture and trade. Several studies of biofuel production processes are assessed in section 3 as to the amounts of energy required and produced as well as economic balances. The assumptions of several authors on this issue are also looked at in this section, where the focus is on ethanol, fuel from oilseeds, and non-food energy crops. Section 4 outlines a set of calculations comparing biofuels with conventional fuels, specifically in terms of ethanol as a gasoline additive and RME (rapeseed oil methyl ester) as a diesel substitute. The CO_2 and other greenhouse gas emissions required to produce biofuels, and emitted during their use, are calculated in section 5 and compared to conventional fuels. Section 6 concentrates on the economics of biofuel production and the cost of using biofuels as a means of reducing CO_2 emissions, as well as the effects of subsidies on biofuel production costs. The final section draws conclusions from all sections as an aid to policy-makers regarding the potential benefits of biofuels and options for the future.

70. Jackson, Marilyn, ed. *Energy and Environmental Strategies for the 1990s.* Lilburn, Ga.: Fairmont Press, 1991.

This volume contains numerous papers written by engineers and energy managers presented at the thirteenth World Energy Engineering Congress. Energy management is increasingly becoming a necessary element in a company's strategic plan, and business people are beginning to be concerned about maintaining energy supplies such as oil, gas, and electricity. Issues such as increasing environmental regulations; the rise of prices of oil, gas, and electricity; and the reduced reliability of electricity are perceived as paramount problems in the efficient operation of businesses. Included are the latest alternatives to increase energy efficiency and reduce operating costs, as well as articles on indoor air quality, CFC reduction, and emission control technologies.

71. Kursunoglu, Behram N., Stephen L. Mintz, and Arnold Perlmutter, eds. *Global Energy Demand in Transition: The New Role of Electricity.* New York: Plenum Press, 1995.

Keywords: general, electricity, energy analysis, generation

This is a collection of papers presented at the conference "Global Energy Demand in Transition: The New Role of Electricity" in 1994. The book explores the technological prospects for, and the long-term availability of, environmentally sound energy sources for generating electricity. The conference had the following objectives: to analyze the increase and diversification in the use of electricity; assess the technological prospects of clean energy sources; examine the role of non-market forces in relation to energy; study the cost impact and national security implications of the increasing demand for energy; provide an analysis of the transition from fossil to electrical transportation; discuss the role of nuclear energy; and consider the best ways to use the plutonium in dismantled warheads. The book should be of use to government in formulating energy policy, as well as to the public for applications in long-term planning.

72. Laughton, Michael. "Combined Heat and Power: Executive Summary." *Applied*

Energy 53 (1996): 227-33.

Keywords: cogeneration, CHP, efficiency, end-use

This article analyzes several aspects related to the present and future implementation of combined heat-and-power (CHP) systems in the UK and its implications for other nations. The emphasis of the article is on the technological, financial, environmental, and institutional aspects related to CHP systems. The author notes that replacing conventional fossil fuel generating facilities and gas boilers by gas-fired CHP units would significantly reduce emissions of several gaseous pollutants and maximize the use of resources. The rationale for this transformation is increasingly favored by a combination of longer-term demand trends for cleaner energy generation technologies and the increasing growth in world population and energy demand.

73. Leach, Gerald. "The Energy Transition." *Energy Policy* (February 1992): 116-23.

Keywords: biomass, developing countries, policy

This article studies the switch within many households of developing countries from traditional biomass fuels to modern sources of energy. The author notes that this transition is directly related to urban size and, within cities, to household income. This is due to the fact that the major constraints on fuel transition are poor access to modern fuels and the high cost of the appropriate appliances. The article provides reasons to encourage intervention by policy-makers to influence the direction and implementation rate of the energy transition. Also included are several examples of previous policy failures.

74. Lovins, Amory B. "Soft Energy Technologies." *Annual Review of Energy* 3 (1978): 477-517.

Keywords: renewable, economics, energy analysis, end-use, environmental

In this classic article, Amory Lovins examines the problems of the traditional approach to energy supply, which consists of expanding readily available (domestic) energy supplies to meet extrapolated homogeneous demands. The article analyzes the increasing economic, environmental, and political costs of this "hard energy path," and proposes the implementation of an end-use based approach. This approach, termed the "soft path," emphasizes an analysis of the tasks for which energy is needed, in order to find the best way to perform each task. A vital part of this approach is based on the introduction of "soft technologies." These technologies, which represent the main thrust of this article, are relatively simple, diverse, and based on renewable sources. They are matched in scale and energy quality to our range of end-use needs. With reference to end-uses, the article examines issues of structure, scale, and technical status, gives comparisons, and also analyzes the debate over soft energy paths.

75. MacDonald, Gordon J. "The Future of Methane as an Energy Resource." *Annual Review of Energy* 15 (1990): 53-83.

Keywords: general, alternative fuels, environmental, transportation

The growing awareness of the detrimental effects of current patterns of fuel use is stimulating the development of alternative options. This article explores some of the environmental advantages related to the use of methane. The author notes that methane may become essential in the future, particularly if the world turns to a hydrogen-based economy. The article analyzes natural gas as a resource base (specifically, how much of it is used throughout the world), environmental considerations related to the atmosphere and stratosphere, and a comparison of methane to other presently used gases. Technologies for enhancing the use of natural gas and current methane uses in transportation and power plants are explored. Methane clathrates are discussed in terms of their properties, origin, thermodynamic stability, energy balance, direct and indirect evidence of natural clathrates, global occurrence of clathrates, estimates of methane stored in clathrates, and the production of methane from clathrate deposits. The article also includes a discussion on the economics of methane production from clathrates and the generation of hydrogen.

76. Ricketts, Jana, ed. *Competitive Energy Management and Environmental Technologies*. Lilburn, Ga.: Fairmont Press, 1995.

Keywords: DSM, efficiency, environmental, planning

These papers were presented at the seventeenth World Energy Engineering Congress. Several authors analyzed a large variety of energy-related topics, and their contributions resulted in a volume that contains ninety chapters divided into eight sections. Section 1 discusses environmental management in terms of various forms of pollution, and issues related to the reporting of information and energy planning. Section 2 examines water resource efficiency; section 3 addresses energy management strategies. Section 4 examines advances in lighting efficiency and applications. Section 5 explores HVAC systems. Section 6 covers competitive power technologies, and section 7 discusses federal energy management programs. The final section examines issues related to demand-side management.

77. Rosenberg, Paul. *The Alternative Energy Handbook*. Lilburn, Ga.: Fairmont Press, 1993.

Keywords: alternative energy, cogeneration, hydroelectric, solar, wind

This book is a useful and practical guide. It provides addresses for more information on several alternative energy technologies and the role they might play in the future. This study, however, is mainly concerned with technologies that can be used now, in various contexts and in different parts of the world. A better knowledge and understanding of alternative forms of energy will hopefully lead to their widespread use, which will likely improve the quality of life for many as well as the quality of the surrounding environment. Chapter 1 considers our energy options and the factors that control the use of these technologies. Chapters 2 and 3 cover the use of solar energy, both thermal and electrical. In chapter 4, the use of storage batteries is discussed, as well as the technologies required to use them effectively. Chapters 5 to 10 cover the technologies of wood and coal heating, passive energy usage, industrial cogeneration, hydropower, small generating systems, and wind power. Chapter 11 looks at future possibilities.

78. Saroff, Lawrence. "Coal Fuel Cycle Externalities Estimates." *Energy Conversion and Management* 37, 6-8 (1996): 1241-46.

Keywords: economics, environmental

Apart from costs such as labor, capital, fuel, and insurance, there are other costs arising from the production of electricity. These are costs that are not passed on to the consumer and remain as externalities. They can be traced to the effects of global warming and the effects on health from emissions of SOx, NOx, and particulates as well as other factors. The aim of this study is an estimation of externalities for the coal fuel cycle. The stages of this cycle are looked at in order to discern which have the greatest effect, and externalities are then estimated. These externalities are measured in economic terms with a cost-benefit analysis. The article also looks at the damage function approach in five steps: new initiatives, impact-pathway analysis and priorities, global warming, and energy security externalities.

79. Sheahan, Richard T. *Alternative Energy Sources: A Strategy Planning Guide.* Rockville, Md.: Aspen Systems Corporation, 1981.

Keywords: alternative energy, cogeneration, economics, environmental, geothermal, hydroelectric, solar, supply and demand, wind

The various alternatives to the more traditional means of energy generation are the central theme of this work. The nature of the energy, new technologies, concerns, marketing potential, economic considerations, and legislative issues are all addressed. An overall picture of the energy future as it stands, as well as a fairly complete outline of alternative energy supply and demand is discussed in the first two chapters. The following eleven chapters deal with the various energy alternatives in sequence. These are: cogeneration, solar energy (heliostats), wind, small-scale hydroelectric power, thermal storage, wood energy, solid waste energy recovery, alcohol fuels, geothermal energy, coal gasification, and peat energy. The last two sections of the book present environmental concerns and financial considerations associated with the use of the previously mentioned alternatives.

80. Shukla, S. K., and P. R. Srivastava, eds. *Environmental Impact Analysis*. New Delhi: Commonwealth Publishers, 1988.

Keywords: alternative energy, conservation, environmental, renewable

This book opens by stating that "energy is the ultimate resource and, at the same time, the ultimate pollutant." It maintains that we have reached our current level of development through our ability to "harness natural flows and accumulations of energy and turn them to human ends." Our need for energy continues to grow, but there are limits to traditional resources. It has therefore become necessary to turn to alternative methods of power supply. Although this book covers various renewable energy sources, it does not take an explicit preventive orientation. Chapter 1 looks at energy in terms of the magnitude and sources of contemporary energy use; growth and change in energy flows; supplies, depletion, and limits of energy resources; energy technology; energy use and conservation; and perspectives on the energy problem. Chapter 2 discusses energy impact analysis in terms of the organization and methodology of an energy impact assessment and energy data and sources. Chapter 3 covers energy sources in terms of historical background, energy use in the United States, prospects for the future, energy resources, and energy conservation.

81. Sterrett, Frances S., ed. *Alternative Fuels and the Environment*. London: Lewis Publishers, 1995.

Keywords: renewable, alternative fuels, geothermal, hydroelectric, ocean, solar, wind

This book is based on a symposium on renewable energy sources and their role in dealing with an eventual energy crisis. Some of the resources examined are derived from the sun's energy which can be harnessed by photochemical reactions, artificial photosynthesis, or photovoltaic electric power generation. Such harnessed energy is then analyzed in terms of the various uses to which it could be put, such as destroying hazardous chemicals, detoxifying wastewater streams, and producing hydrogen for use as a fuel. The feasibility of generating electricity through wind turbines, oceanthermal, and geothermal resources is considered. Hydroelectricity, as well as biofuels and biomass is also discussed in some detail. Oxygenated fuels such as ethanol and methanol are analyzed with regard to reformulated gasolines. Case studies on the "organic carbonyl compounds in Albuquerque, New Mexico, air," "the effects of oxygenated fuels on the atmospheric concentrations of carbon monoxide and aldehydes in Colorado," "modeling the effects of alternative fuels on ozone in Ontario, Canada," "mitigation of environmental impacts at hydroelectric power plants in the United States," "wind energy in the United States," and "areal wind resources assessment of the United States" add to the value of this book.

Renewable Energy Sources

Almost all the entries in this section have considerable policy implications. This is because the move to renewables is usually initiated in any country by governments or through government subsidies. However, if the main theme of a work was policy initiatives for renewables, it has been placed in the section headed *Energy Policy for Sustainability* in the bibliography.

82. Acker, Richard H., and Daniel M. Kammen. "The Quiet (Energy) Revolution: Analysing the Dissemination of Photovoltaic Power Systems in Kenya." *Energy Policy* 24, 1 (1996): 81-111.

Keywords: renewable, case studies, policy, energy analysis, developing countries, photovoltaics

Photovoltaic electricity has a fairly long history in terms of experimental and small-scale applications; however, its widespread adoption still remains far from common. This article documents and examines the emergence and growing use of small-scale photovoltaic (PV) systems in Kenya. During the past decade, between 20,000 to 40,000 PV systems were installed in Kenya. This impressive record has consisted of mostly privately financed installations. The analysis provided in this article offers valuable lessons for understanding the development and spread of renewable energy technologies. It also illustrates the actual and potential roles of different stakeholders such as international development organizations, grassroots organizations, and private entrepreneurs.

83. American Solar Energy Society. *1990 ASES Roundtable: Renewable Energy Options for Utilities: Developing Partnerships*. Boulder, Colo.: American Solar Energy Society, 1990.

Keywords: renewable, policy, solar

This book contains the proceedings of the 1990 American Solar Energy Society (ASES) Roundtable on renewable energy options for utilities. Energy experts, environmentalists, and stakeholders from the energy industries participated in the conference. The main goal of this volume is to "show a way to use clean, renewable energy for electric power generation." The proceedings are organized into three main sections. The first, "Solar Energy Technologies," discusses current technologies in renewable energy. Section 2, "Policy and Regulatory Issues," examines policy issues that concern renewables. Section 3, "Forging Partnerships," covers methods to ensure the success of projects, with a focus on the utility industry. After each of these sections there is a panel discussion, and an "Open Forum" where the audience participated in a Roundtable. Following this section are concluding remarks by several speakers.

84. Anderson, Dennis, and Ahmed Kulsum. *The Case for Solar Energy Investments*. Washington, D.C.: The World Bank, 1995.

Keywords: alternative energy, biomass, economics, photovoltaics, solar, wind

As problems associated with conventional forms of energy generation become widely recognized, alternative methods of energy production are being sought. Within the World Bank's energy series, the status and costs of renewable energy technologies have been reviewed. This particular report discusses four technologies: photovoltaics, solar-thermal, wind, and biomass, and argues that their widespread development should be strongly encouraged. These technologies are favored by the following facts: solar energy is abundant; there has been significant progress in lowering costs and improving operational performance; economic prospects are very positive; and environmental preservation has become a pressing policy need. Based on these reasons the report proposes a concerted international initiative to accelerate the commercialization of solar technologies. A two-part program is outlined to help with this commercialization in developing countries. First, a "pipeline" of investments needs to be established to draw financial resources. Secondly, public research and development at the national and international levels in support of private initiative needs to be expanded.

85. Berman, Daniel M., and John T. O'Connor. *Who Owns the Sun? People, Politics, and the Struggle for a Solar Economy*. White River Junction, Vt.: Chelsea Green Publishing Company, 1996.

Keywords: renewable, policy, economics, sustainability, solar, photovoltaics, planning, cities, case studies, electricity, reliability

The connections between energy and democracy are unfortunately rarely explored in much detail. This lack of debate presents a problem for renewable energy technologies, which are inherently suited for decentralized power generation. Therefore, the discussion presented in this volume on the shape and form of ownership of future energy developments is extremely relevant and necessary today. The main thesis of *Who Owns the Sun?* is that "local ownership and democratic control of energy are the necessary, if not sufficient, conditions for a solar economy." The authors believe that municipal ownership of electric companies and citizen electric cooperatives constitute appropriate models for energy administration. To illustrate their arguments, Berman and O'Connor analyze the historical development of the public-power movement in the United States. Their work also thoroughly analyzes the political and economic maneuvers of different stakeholders of the energy business in order to explain the current domination of the energy field by the fossil fuel industry. Although this volume focuses mostly on the energy situation of the United States, the ideas presented by the authors can be tailored to understand energy issues in many other regions of the world.

86. Blackburn, John O. *The Renewable Energy Alternative: How the World Can Prosper Without Nuclear Energy or Coal.* Durham, N.C.: Duke University Press, 1987.

Keywords: alternative energy, case studies, conservation, efficiency, sustainability

This book was written for those who believe that a continued reliance on sources of energy such as coal and nuclear power is not a satisfactory means of producing the energy we need. It examines numerous progressive utilities which are embracing the use of energy alternatives. Designed as a tool for those in energy supply industries, in government, or in private agencies concerned with energy policy, this book brings together the strands of energy analysis of Denis Hayes, Amory Lovins, and the Union of Concerned Scientists. It discusses step-by-step how to get to a renewable and sustainable energy future by the year 2000, despite having been initiated in 1977. The book begins by examining the energy debate and illustrates a severe lack of agreement among professionals. It then provides its view on energy-conservation-efficiency potential and how it could be implemented within the United States, resulting in reductions in coal, synfuels, and nuclear power. Other industrialized nations are examined as case studies, and the prospect of such a transition is assessed for the Third World. The book concludes with policy proposals regarding costs, research, and incentives.

87. Bockris, John O'M., T. Nejat Veziroglu, and Debbi Smith. *Solar Hydrogen Energy: The Power to Save the Earth.* London: Macdonald & Co., 1991.

Keywords: alternative energy, solar hydrogen

Evidence of the detrimental effects of using fossil fuels to power our civilization can be found in problems such as global warming, acid rain, pollution, and ozone depletion. This book considers the generation of energy using solar-hydrogen systems as a means to avoid many of the negative effects created by fossil fuels. The authors note that this energy option is a safe, clean, and sustainable alternative. It works by harnessing sunshine to generate electricity and using that electricity to electrolyze water into its two main components: hydrogen and oxygen. Hydrogen is a highly efficient and versatile fuel. The authors believe that hydrogen could be used to produce electricity and heat, as well as to fuel cars, aircraft, ships, and trains. The authors maintain that it is not necessary to continue using damaging carbon-containing fuels when there is an abundant alternative waiting to be exploited.

88. Boyle, Godfrey. *Renewable Energy: Power for a Sustainable Future*. Oxford: Oxford University Press, 1996.

Keywords: economics, environmental, fuels, renewable

This excellent reference volume covers numerous aspects of energy alternatives. It is filled with charts, graphs, tables, maps, and pictures, which provide abundant information and make the volume quite easy to read. Chapter 1 gives a general overview of the entire book, covering present-day fuel use, the energy problems of modern societies, and a brief examination of the various renewable energy sources available to us. Chapters 2 through 9 provide an in-depth explanation of each of the energy alternatives explored in chapter 1, covering everything from the science of the various forms of each technology to their economic, environmental, and social benefits and costs. The various alternatives examined include: solar thermal energy, solar photovoltaics, biomass, hydroelectricity, tidal power, wind energy, wave energy, and geothermal energy. Chapter 10 deals with the means of integrating these alternatives into society and covers the economics of transition, promotion, sources, and long-term studies. An appendix is also included which examines topics involving cost and resource estimation. The overall approach of this book is interdisciplinary, covering the economic, social, environmental, and policy issues raised by renewable energy as well as the physical and engineering aspects.

89. Brower, Michael. *Cool Energy: The Renewable Solution to Global Warming*. Cambridge, Mass.: Union of Concerned Scientists, 1990.

Keywords: environment, fuels, policy, renewable

It is becoming more and more evident that we cannot continue using fossil fuels at current rates due to their harmful effects on the environment. This realization acts as an incentive to find and develop alternative means of energy generation. Brower examines fossil fuels and global warming, energy strategies, and various types of renewable energy, such as solar energy, wind energy, biomass energy, and hydroelectric power. For each of these renewable energy sources, Brower looks at the resource itself, technologies and economics, storage, and environmental impacts. The final chapter discusses market barriers and policy recommendations on the "path to a renewable future."

90. Brower, Michael. *Cool Energy: Renewable Solutions to Environmental Problems*. Cambridge, Mass.: MIT Press, 1992.

Keywords: renewable, biomass, economics, fuels, geothermal, hydroelectric, ocean, solar, wind

Brower takes an in-depth look at renewable energy technologies such as solar power, wind energy, biomass energy, hydroelectric power, and geothermal energy as a way out of an escalating energy crisis. While the consumption of energy in countries like the United States is growing, people are demanding that energy be produced from clean, safe sources. Brower believes that it is desirable and practical to move from using fossil fuels to "cool" renewable energy. The popularity of renewable energy has faced many obstacles in the past under different government officials and business leaders, as well as from poorly conceived implementation projects. It is, however, gaining a reputation for being more and more reliable, efficient, and cost-effective. Chapter 1 looks at some "economic and environmental consequences of America's fossil-fuel-based economy" demonstrating that reducing such methods of energy generation will prove beneficial in both areas. Chapter 2 describes how renewable energy can provide us with most of our energy needs and outlines its advantages and potential compared to fossil fuels and nuclear power, as well as the barriers that stand in the way of its growth. Chapters 3 through 8 look at solar energy, wind energy, biomass, energy from rivers and oceans, geothermal energy, and energy storage respectively. The last chapter, chapter 9, discusses policies for a sustainable future.

91. Butti, Ken, and John Perlin. *A Golden Thread: 2500 Years of Solar Architecture and Technology*. New York: Van Nostrand Reinhold, 1980.

Keywords: renewable, case studies, solar, energy efficient buildings, cities, planning

The use of solar energy to heat dwellings dates back at least 2500 years, to the ancient Greeks. This volume documents the evolution of solar architecture in the Western world and also contains some references to similar achievements in pre-Columbian America, China, and Japan. Developments in solar architecture and technology are discussed within their historical, socioeconomic and political contexts. This elegantly written book supplies abundant historical examples that provide creative ideas and inspiration, and is a valuable resource for readers interested in renewable energy issues.

92. Carless, Jennifer. *Renewable Energy: A Concise Guide to Green Alternatives*. New York: Walker, 1993.

Keywords: alternative energy, biomass, economics, fuels, geothermal, hydroelectric, renewable, solar, wind

The struggle for the widespread use of alternative forms of energy generation is ongoing. This book provides a brief account of various forms of renewable energy, including solar, wind, hydropower, geothermal, and biomass, as well as renewable automobile fuels. The author also analyzes energy efficiency and energy prospects for the future. All the alternative forms of energy generation presented are discussed in terms of their history, current status, benefits and concerns, different uses, costs, and future prospects. The author attempts to demonstrate that, contrary to popular belief, renewable energy is ready to be used on a much larger scale today. Her goal is to prove that the technical difficulties have been greatly reduced and that the remaining barriers are mostly political.

93. Carlson, D. E. "Photovoltaic Technologies for Commercial Power Generation." *Annual Review of Energy* 15 (1990): 85-98.

Keywords: renewable, photovoltaics, solar

In the last decade, solar cells have found widespread use in numerous applications such as power sources for microwave repeater stations, remote telemetry systems, cathodic protection systems for pipelines and bridges, and solar-powered calculators. As the cost of photovoltaic systems continues to decrease, applications such as water pumping and power for remote villages are starting to become cost-effective. Moreover, with improvements in performance and continued reductions in the manufacturing cost, photovoltaic power generation should start to penetrate the utility grid market by the turn of the century. If low-cost energy storage systems become available, photovoltaics could become one of the major sources of energy in the next century. This paper reviews the status of various photovoltaic technologies as well as present applications. The prospects for both distributed and central station grid-connected systems are discussed. The paper concludes with a discussion of the institutional and political factors that will affect the introduction of grid-connected photovoltaic power systems.

94. Coldicutt, Susan, and Terry Williamson. "Concepts of Solar Energy Use for Climate Control in Buildings." *Energy Policy* (September 1992): 825-35.

Keywords: renewable, energy efficient buildings, solar

This paper deals with solar energy for heat generation. The authors stress that the term "solar energy use" is very broad and needs to be well defined and "qualified regarding purpose of quantification, type of energy, definition of use, baselines and context." Concepts of energy used in buildings are analyzed in terms of various forms of solar energy use for passive heating in buildings, and the use of physics to "describe and analyze the solar energy reaching the earth and that used in buildings." Concepts of people's relations to buildings are discussed in an effort to quantify energy use and as a means to understand how we alter the environment to increase our comfort. The article then addresses the term "solar energy use" in greater depth and notes that the use of solar energy does not lessen the total amount available. Finally, the article examines the issues of how the theories can and should be applied, as well as the implications involved.

95. Commission of the European Communities. *The European Renewable Energy Study.* Brussels: Office for Official Publications of the European Communities, 1994.

Keywords: renewable, Europe, case studies, consumption, economics, electricity, policy

This report consists of four volumes organized as follows: Main Report (volume 1); Technology Profiles (volume 2); Country Profiles (volume 3); Reference Data (volume 4). The main goal of the report was to evaluate the long-term prospects for renewable energy technologies within European Community nations and countries of Central and Eastern Europe. The study analyzes the current technical performance of various renewable energy technologies; the costs of existing technologies at current prices; the prospects for developing the technologies; and potential cost reductions related to technological progress and mass production. The report also determines the technical and economic potential of each technology; assesses the possible penetration of technologies up to 2010 based on four different scenarios, and provides an analysis of the development constraints of renewable energy technologies and strategies to overcome these constraints.

96. Cook, Jeffrey. *Award-Winning Passive Solar House Designs*. Charlotte, Vt.: Garden Way Publishing, 1984.

Keywords: renewable, energy efficient buildings, solar

Cook examines a large number of different solar house designs, all chosen from among award winners and finalists from the First Passive Solar Design Awards Competition. The designs were chosen from among 350 entries from several countries, by a "distinguished jury of architects, engineers and other professionals" which selected the designs in this book for their "excellence in the synthesis of architecture and engineering." The houses presented in this volume are organized "according to themes based on climate, and construction or building type." This volume will be of interest to readers who wish to further their knowledge of solar architecture and constitutes a concise resource for various creative and unique designs.

97. Dmeljanousky, Erazm, Robert Hill, and Ivan Chambouleyron, eds. *Advanced Technology Assessment System Prospects for Photovoltaics: Commercialization, Mass Production and Application for Development*. New York: United Nations, 1992.

Keywords: renewable, case studies, developing countries, photovoltaics, policy

The largest portion of this book originates from papers presented at the International Workshop on Photovoltaics held in Sao Paulo, Brazil in 1991. In order to achieve a more comprehensive collection, additional works by various experts on the subject were added. The papers deal with the widespread dissemination of photovoltaic (PV) equipment, especially in the Third World; PV technology and production; policy aspects of PV distribution; and the experience of several developing countries in the production and use of PV systems. Chapter 1 contains papers that provide an overview on PV cell and module development, the production of PV equipment, and the role of PV electricity in world energy supply. Chapter 2 covers issues related to PV technology and its costs. Chapter 3 discusses policy issues, and Chapter 4 includes reports from Third World countries on their experience with PV production and implementation.

98. El-Hinnawi, Essam, and Asit K. Biswas. *Renewable Sources of Energy and the Environment*. Dublin: Tycooly International, 1981.

Keywords: renewable, environment

During the 1970s the general realization that fossil fuels, especially oil and natural gas, are finite in nature brought to the forefront the need for energy alternatives. Like most sources of energy, renewables have a number of interrelated environmental effects within nations and international regions. This book attempts to give a balanced assessment of these effects and encourages the scientific community to find adequate solutions to the problems encountered, or likely to be encountered. Topics covered include: opportunities for renewable sources of energy; geothermal, solar, wind, hydroelectric, and biomass energy sources; and energy storage systems.

99. Gilchrist, Gauin. *The Big Switch: Clean Energy for the Twenty-first Century*. Vol. 1. St. Leonards, Australia: Allen & Unwin, 1994.

Keywords: renewable, economics, efficiency, electricity, environmental

Governments and the electricity industry seem to be of the opinion that implementing solar and wind energy would increase energy costs. This book, however, argues that the opposite is actually true and that we are losing money through the continued use of obsolete technologies for energy production. The author maintains that "governments have conspired with their power companies to stifle the promotion of new high-technology energy-efficient systems because they have become addicted to using these companies as tax collectors; as a result, power bills are higher than they should be and Australian industry is less competitive internationally." This book addresses a major contemporary issue: "How can modern societies meet their essential electricity needs while reducing their impact on the environment?" The belief is that if Australia makes "the Big Switch," it could have an electricity industry that pollutes less, costs businesses and households less, and creates more jobs.

100. Gipe, Paul. *Wind Energy Comes of Age*. New York: John Wiley & Sons, 1990.

Keywords: renewable, case studies, economics, environment, solar, wind

This volume evaluates the present and future status of wind energy. It discusses the ups and downs of the California and Danish markets, as well as wind power technology in terms of economics and reliability. Denmark is praised for the progress it has made in wind energy technology and because it stays away from more sophisticated but less reliable designs. The author notes that wind power has become cheaper than coal and nuclear power for many areas of the world. However, he warns against extrapolating wind energy's falling costs indefinitely and to avoid claiming that this source will become "too cheap to meter." Wind technology has faced some problems, such as bird kills, faulty government sponsored research and development, and the use of wind machines that are too large and unreliable. Part 1 evaluates present wind technology achievements and possible future technological advancements. Part 2 covers some of the detrimental impacts of wind energy in terms of aesthetics, people, land, and wildlife; and how these problems can be solved. This part also examines the positive aspects of wind technology and how it can grow to its full potential, and remain an environmentally sound method of energy production for the future. Part 3 looks at wind turbines and their integration with utilities, and possible alternative generation combinations using wind energy and methane. Also explored are ways to avoid rapid cycles of boom and bust, such as those experienced in California during the 1980s, by implementing strategies that encourage the "sustained and orderly development" of wind projects.

101. Gipe, Paul. "Wind Energy's Declining Costs." *Solar Today* 9, 6 (1995): 22-25.

Keywords: renewable, economics, planning, wind

The cost of energy produced by wind turbines has become competitive with that of more predominant sources of energy. Although this fact constitutes a great success story, this article warns that some commentators are beginning to reach absurd conclusions about wind energy. Ken Karas, who was president of the American Wind Energy Association (AWEA), criticizes the methods used to reach those kinds of conclusions. As an example he cites the use of the COE (cost of energy) method of calculating costs. Another argument made is that two distinctly different concepts, the installation cost and the installation price, are often confused. The cost of operating and maintaining wind turbines is an additional factor influencing the cost of energy. The article notes the effect of installed cost on the cost of energy and explores future costs for energy generated by wind turbines. The author states that, frequently, cost comparisons between different forms of energy generation overlook factors that are difficult to quantify in terms of price. The article also discusses the decommissioning of wind plants; the flexibility and cost savings that modularity offers to utility planners; social and environmental costs; the cost of capital; and finally, bidding and price.

102. Golob, Richard, and Eric Brus. *The Almanac of Renewable Energy*. New York: Henry Holt, 1993.

Keywords: renewable, solar, wind, nuclear, hydroelectric, biomass, geothermal, photovoltaics, environmental impacts

This is a useful introduction to, and evaluation of, several renewable energy technologies. The first chapter discusses the current dominant sources of energy generation—coal, oil, natural gas, and nuclear generation—as a prelude for a discussion of renewable sources. As part of that discussion, seven chapters deal with hydroelectric, biomass, geothermal, solar thermal, photovoltaic, wind, and ocean energy. Two concluding chapters, one on energy storage and another on energy efficiency, advocate a strategy to increase the use of renewable energy at a global level. Each chapter also describes the technical means related to each energy source, their environmental costs and benefits, and their current scope and future potential. The authors have included two useful appendices. Appendix A contains a brief section on energy definitions, units, and conversion tables, as well as numerous tables containing detailed information on regional patterns of energy consumption, distribution of energy resources, governmental funding levels, and descriptions of several energy facilities. Appendix B provides a limited list of additional information sources on renewable energy and energy efficiency.

103. Green, Joanta. *Renewable Energy Systems in Southeast Asia*. Tulsa, Okla.: Pennwell Books, 1996.

Keywords: renewable, case studies, photovoltaics, policy, developing countries

The implementation of renewable energy projects requires an analysis of regional energy markets. This volume surveys the market prospects of nonconventional power generating and transforming equipment in the Asia-Pacific Region. As part of this survey, the author analyzes the energy economy of the region, as well as trends in economic growth, electricity use, energy intensity, and investment. The energy profiles and renewable energy initiatives of Indonesia, Malaysia, the Philippines, and Thailand are analyzed in detail. The focus of the renewable energy analysis is on solar photovoltaic systems, small hydropower, wind energy, solar thermal energy, and biomass (including biogas digesters, biomass gasifiers, and combustors). The author observes that these technologies have often been handled by agencies in charge of rural development which have achieved results below the systems' potential. One of the main suggestions of this volume is that renewable energy sources that exploit conventional energy generation equipment should be actively marketed in the region.

104. Headley, Oliver. "Renewable Energy Technologies in the Caribbean." *Solar Energy* 59, 1-3 (1997): 1-9.

Keywords: case studies, photovoltaics, renewable, wind

Practicality has been one of the main incentives for the use of renewable energy on small islands, and this article examines how energy needs can be met through the use of such systems, using the Caribbean as a case study. Renewable energy is nothing new to the Caribbean, as windmills were once used to grind sugarcane. Wind energy for these islands as a source of power makes perfect sense as wind speeds are in excess of 10 m/s, and interest in this source of energy for electricity as well as other uses is quickly growing. The five main topics discussed are: solar water heating, solar drying, solar distillation, solar cooling, and photovoltaic power.

105. Howes, Ruth, and Anthony Fainberg, eds. *The Energy Sourcebook: A Guide to Technology, Resources, and Policy*. New York: American Institute of Physics, 1991.

Keywords: renewable, solar, policy, photovoltaics, ocean, wave, biomass, energy efficient buildings, hydroelectric, efficiency, geothermal, nuclear

The purpose of this sourcebook is to provide a detailed introduction to several of the alternative sources available for energy generation. The authors describe, and analyze in detail, the scientific, economic, and environmental aspects related to these alternative energy sources. The volume starts with an analysis of energy issues since the 1973 energy crisis and the risks associated with energy production, followed by a presentation of several sources of energy including: fossil fuels; nuclear power; fusion; photovoltaics; solar thermal; hydroelectricity; geothermal energy; energy from the oceans; and biomass. The final section of the volume analyzes several end-use technologies including: energy storage systems; transportation; agriculture; manufacturing; buildings and appliances; and the impact of materials technologies on motor-drive systems.

106. Johansson, Thomas B., ed. *Renewable Energy: Sources for Fuels and Electricity*. Washington, D.C.: Island Press, 1993.

Keywords: renewable, case studies, economy, electricity, environment, fuels, hydroelectric, photovoltaics, solar, wind

This volume is the result of an international cooperation effort among some of the most renowned renewable energy experts, supported by the governments of Sweden, Norway, and The Netherlands. Its mandate was to generate an accurate report on the latest advances in renewable forms of electricity and fuel generation from a technical and economic perspective, to be used as background information for the United Nations Conference on Environment & Development (UNCED). The following topics are analyzed: renewable fuels and electricity for a growing world economy; hydropower and its constraints; wind energy in terms of technology, economics, and resources; energy systems and regional strategies; solar-thermal electric technology; photovoltaic technology, including crystalline and polycrystalline-silicon solar cells, photovoltaic concentrator technology, amorphous silicon photovoltaic systems and polycrystalline thin-film photovoltaic systems; ocean energy systems; geothermal energy; biomass for energy including bioenergy, open-top wood gasifiers, advanced gasification-based power generation, biogas electricity (a case study), anaerobic digestion for energy production and environmental protection, the Brazilian fuel-alcohol program, and the production of ethanol and methanol from cellulosic biomass; solar hydrogen; utility strategies for using renewables; and a global renewable energy scenario.

107. Kimani, Muiruri J., and Ekkerhart Naumann, eds. *Recent Experiences in Research, Development and Dissemination of Renewable Energy Technologies in Sub-Saharan Africa*. Nairobi: Kengo, 1993.

Keywords: renewable, developing countries, case studies, policy, economics

This is the proceedings of a regional seminar held in Nairobi, Kenya, in 1993. This seminar reviewed field experiences in the research, development, and dissemination of renewable energy technologies in Sub-Saharan Africa. The proceedings involved a regional renewable energy technology review, followed by an analysis of the economic viability of these technologies and the presentation of several renewable energy case studies implemented in Kenya, Ethiopia, Sudan, Tanzania, and Uganda. Political, economic, and other issues that constitute obstacles for the dissemination of renewable energy strategies in the region are discussed. The final section evaluates a M.Sc. course on renewable energy offered by Oldenburg University. This section may be of interest to readers interested in renewable energy education programs

108. Kishore, V. V. N., and N. K. Bansal. *Renewable Energy for Rural Development: Proceedings of the National Solar Energy Convention*. New Delhi: McGraw-Hill, 1989.

Keywords: renewable, case studies, biomass, generation, photovoltaics, solar, wind

This book covers the proceedings of the twelfth annual convention of the Solar Energy Society of India (SESI) held in 1988. The main theme of the convention was renewable energy for rural development. Papers presented at the conference were written by researchers, entrepreneurs, and policy-makers concerned with renewable energy technologies. The availability of electricity is presently very low in rural India, although slow progress in electrification is being made. The book's objective is to act as a catalyst to increase the availability of electricity through renewable energy means. Part 1, "Science and Technology," examines various forms of renewable energy such as solar thermal, solar photovoltaics, wind energy, biomass energy, passive solar, and others. Part 2, "System and Applications," discusses energy systems and is concerned with areas such as cooking, water pumping, and power generation. Part 3 presents case studies concerning renewable energy.

109. Kulsum, Ahmed. *Renewable Energy Technologies: A Review of the Status and Costs of Selected Technologies*. Washington, D.C.: The World Bank, 1994.

Keywords: renewable, biomass, economics, fuels, photovoltaics, solar

This volume examines photovoltaics, solar thermal applications, and biomass for liquid fuels and electricity production, in terms of their past and future economic costs. The author also discusses the development of these renewable energy applications over time, and notes that decreases in cost will continue with technical progress and market growth. His goal is to assist in the development of cost-effective technologies that will reduce emissions of carbon dioxide. The study demonstrates an overall decrease in the costs of renewable forms of energy generation. It also notes some of the limitations of renewable energy sources, such as the facts that electricity produced from biomass is quite site specific and that electricity production by solar thermal sources is still largely at an experimental stage.

110. Larson, Ronal W., and Ronald E. West, eds. *Implementation of Solar Thermal Technology*. Vol. 10. Cambridge, Mass.: MIT Press, 1996.

Keywords: energy efficient buildings, environmental, policy, solar

The twelve volumes in this series cover the research, development, and implementation of solar thermal energy technologies that were carried out during the last eleven years of the National Solar Energy Program. This particular book addresses the programmatic aspects of government involvement in solar energy and assesses the part of the government solar energy program that was commercialized, which caused it to be controversial at first. Part 1 of this volume deals with the implementation of solar thermal technologies, and contains an introduction and papers on the role of Congress as well as market development. Part 2 explores solar thermal program perspectives in terms of active heating and cooling, passive technologies, passive commercial building activities, industrial process heat, and high-temperature technologies. Solar thermal demonstrations and construction are treated in part 3, and papers in this section discuss residential, commercial and federal buildings, agricultural demonstration programs, and military demonstration programs. Part 4 reviews consumer assurance and testing, standards, and certification of solar thermal quality assurance. Solar thermal information is addressed in part 5, with papers on consumer information, public information, technical information, training and education, and regional solar energy centers. Part 6 discusses solar thermal technology transfer in terms of its liaison with industry, the Solar Energy Research Institute, the Los Alamos National Laboratory, and the Argonne National Laboratory. Solar thermal incentives are looked at in part 7, specifically in terms of tax credits, financing, and grants. The last part of this volume deals with solar thermal organizational support with papers on the following topics: international activities, state and local programs, and public utilities, as well as legal, environmental, and labor issues.

111. Liebenthal, Andres, Subodh Mathur, and Herbert Wade. *Solar Energy: Lessons from the Pacific Island Experience*. World Bank Energy Series. Vol. 244. Washington, D.C.: World Bank, 1994.

Keywords: renewable, photovoltaics, developing countries, case studies, policy, economics, electricity

The vast majority of the population of developing nations living in rural areas have very limited access to electricity. This report analyzes the experience of Pacific island countries in rural electrification using photovoltaic (PV) equipment. The authors suggest that one of the most promising approaches for the provision of PV-based electricity in remote areas is one based on fee-for-service, administered by local, cooperatively owned utilities. In their opinion, this system is superior to those based on the sale of hardware. They note that the success of PV solar programs depends on the choice of adequate technology, proper training of local maintenance personnel, effective fee collection systems, and sound financial administration. The report also contains a life-cycle cost analysis that shows that the costs of PV systems are marginally lower than those of diesel systems for households in remote rural areas.

112. MacGregor, Kerr, and Colin Porteous, ed. *North Sun 94: Solar Energy at High Latitudes*. London: James and James, 1994.

Keywords: renewable, energy efficient buildings, solar, photovoltaics, pollution prevention, efficiency, reliability

This is a compendium of papers presented at the 1994 North Sun Conference in Glasgow, Scotland. The theme of the North Sun Conferences is solar energy applications at high latitudes. The papers are divided into the following areas: solar water heating; active solar heating; photovoltaic applications; solar modeling and design tools; solar buildings (including sun spaces and greenhouses); solar preheat for ventilation and solar insulation; windows and solar lighting; solar policy and implementation strategies. This volume will be of interest to both solar energy experts and readers who seek to increase their technical knowledge of solar applications.

113. McVeigh, James, Dallas Burtraw, Joel Darmstadter, and Karen Palmer. "Winner, Loser, or Innocent Victim? Has Renewable Energy Performed as Expected?" *Solar Energy* 68, 3 (March 2000): 237-55.

Keywords: renewable, biomass, geothermal, alternative fuels, alternative energies, photovoltaics, electricity

This study provides an evaluation of the performance of five renewable energy technologies used to generate electricity—biomass, geothermal, solar photovoltaic, solar thermal, and wind. The authors compare the actual performance of these technologies against stated projections that helped shape public policy goals over the last three decades. Their findings document a significant difference between the actual and predicted success of renewable technologies in penetrating the U.S. electricity generation market and in meeting cost-related goals. In general, renewable technologies have failed to meet expectations with respect to market penetration. They have succeeded, however, in meeting or exceeding expectations with respect to their cost. Their failure at market penetration is attributable to the cost of conventional generation having fallen as well, mostly because of entrenched subsidies. Obviously, if the full costs of conventional electricity generation were internalized, renewable energy technologies would then become an economically lucrative option.

114. Miller, Damian, and Chris Hope, "Learning to Lend for Off-grid Solar Power: Policy Lessons from World Bank Loans to India, Indonesia, and Sri Lanka." *Energy Policy* 28, 2 (February 2000): 88-93.

Keywords: alternative energy, policy, solar, photovoltaics, developing countries

The World Bank has sought to advance the diffusion of solar photovoltaic technology for off-grid applications in the developing world. As these systems are fundamentally different from centralized power stations and conventional rural electrification, the World Bank has been learning how best to lend for such technology. This study seeks to highlight the lessons learnt from the first such loans to India, Indonesia, and Sri Lanka. It uses lifetime cost analysis to justify continued interventions in this sector, and draws on theories of innovation diffusion to guide analysis, and ultimately, policy recommendations

115. Ogden, Joan M., and Robert H. Williams. *Solar Hydrogen: Moving Beyond Fossil Fuels*. Washington, D.C.: World Resources Institute, 1989.

Keywords: renewable, alternative fuels, environmental, photovoltaics, solar

This book is based on the premise that the energy crisis of the 1970s has reemerged, albeit in a more complex form than before. This is due to the fact that today numerous environmental problems complicate the realm of energy. The use of solar hydrogen, as a means of producing power, is seen as a way to alleviate concerns about the scarcity of energy and environmental problems. The authors note that hydrogen deserves a closer look because of advances in photovoltaic (PV) cells, which directly convert sunlight into electricity. This electricity can then be used to electrolyze water and produce hydrogen, a clean burning fuel. This technology is very close to being cost competitive, which will influence the development and acceptance of new products such as hydrogen-powered cars. The authors believe that hydrogen fuel could replace fossil fuels soon after the turn of the century, a change that will have significant economic and environmental benefits, and that changes to public policy are imperative to implement a PV-hydrogen system. The following subjects are covered: various forms of pollution, the silicon solar cell, designing a PV hydrogen energy system, how PV hydrogen compares to other synthetic fuels, how it can replace oil, breaking into the markets, and solar hydrogen for transportation.

116. Pai, B. R., and M. S. Ramaprasad, eds. *Power Generation Through Renewable Sources of Energy*. New Delhi: McGraw-Hill, 1991.

Keywords: renewable, case studies, policy, photovoltaics, tidal, biomass, developing countries, wind, economics

The increasing gap between the demand and supply of energy requires that new and renewable sources of energy generation and conservation should be constantly developed and implemented. Several Indian organizations have made considerable efforts towards the generation of renewable energy sources in the past ten to fifteen years. Some of these efforts are documented in this volume, which contains papers presented in a seminar held at the Indian Institute of Science in Bangalore on November 28-29, 1989. The main focus of the seminar was on the techno-economic considerations of power generation using solar photovoltaic systems, wind power, biogas, and wave energy.

117. Ravindranath, N. H., and D. O. Hall. *Biomass, Energy, and the Environment.* Oxford: Oxford University Press, 1995.

Keywords: alternative energy, biomass, developing countries, environmental, sustainability

This volume supports the view that the use of biomass energy, in developing as well as industrialized countries, will likely become crucial in working towards sustainable development. In a world where environmental problems related to energy use are constantly increasing, the potential of biomass for energy generation must be closely scrutinized. The authors provide a detailed analysis of the use of biomass energy in India. India currently uses a large range of biomass resources and has implemented many programs that have had varying degrees of success. Bioenergy is compared to other renewable energy options by exploring its development potential, socioeconomic issues, and environmental impact. Policy options to promote the use of bioenergy are provided. The authors conclude that several developing countries in Sub-Saharan Africa, South-East Asia, and Latin America, which are in a similar position to India in terms of high population densities and fairly rich biomass resources, can also benefit from new biomass technologies. The authors use a "bottom up" approach to describe and outline policy issues for each country or region.

118. Reinhard, Haas. "The Value of Photovoltaic Electricity for Society." *Solar Energy* 54 (January 1995): 25-31.

Keywords: renewable, conservation, environmental, photovoltaics

This article weighs the benefits of photovoltaic (PV) systems for society by examining them from different perspectives: consumer, utility, and environmental, as well as in relation to central and decentralized systems. It looks at several issues to be addressed when considering the use of PV, and takes into account feedback from consumers who draw electricity directly from the sun. The article sees changes in consumer behavior in terms of decentralized PV systems as triggering energy conservation and load-shift effects. It is proposed that the collective benefits of PV systems are responsible for its growing use as an alternative source of energy which will eventually penetrate the market.

119. Saha, H., S. K. Saha, and M. K. Mukherjee, eds. *Integrated Renewable Energy for Rural Development: Proceedings of the National Solar Energy Convention.* New Delhi: McGraw-Hill, 1990.

Keywords: renewable, alternative energy, poverty alleviation

This book contains the proceedings of the fourteenth annual convention of the Solar Energy Society of India. The theme was "Integrated Renewable Energy for Rural Development." One hundred and forty-four papers were selected for the conference, and were divided into the following sections: solar thermal, solar photovoltaics, bio-mass energy, wind, tidal and geothermal energy, chemical, photochemical and other routes, energy planning, and integrated rural energy. The solar thermal section deals with several subjects: low temperature applications, high temperature applications and materials, cooling, and other applications. Several papers also deal with the following issues on photovoltaics: silicon solar cells, compound semiconductor solar cells, and "systems and applications."

120. Shashua-Bar, L., and M. E. Hoffman. "Vegetation as a Climactic Component in the Design of an Urban Street." *Energy and Buildings* 31, 3 (April 2000): 221-36.

Keywords: cities, policy, case study

The cooling effect of small urban, green, wooded sites of various geometric configurations is the subject of this study. The authors perform a statistical analysis of 714 experimental observations from eleven wooded sites in Tel Aviv. The results demonstrate that the wooded sites had a noticeable cooling effect not only on structures in their immediate vicinity but also on buildings as far as 100 meters away. The empirical findings of this study permit the development of tools for incorporating the climactic effects of green areas into urban design. Some policy measures are indicated for alleviating the "heat-island" effect in the urban environment.

121. Sodha, M. S., N. K. Bansal, and M. A. S. Malik. *Solar Passive Building.* Oxford: Pergamon Press, 1986.

Keywords: renewable, energy efficient buildings, solar

This book covers the fundamentals of the science and architecture of solar passive building, showing how optimal mass insulation, orientation, shape, built form, and layout play their roles in the design process. Starting from the known thermal indices, defined to satisfy the comfort level in the given space, the book describes many parameters and concepts relevant to the design and study of solar passive buildings. Topics covered include thermal comfort, solar radiation, building clusters and solar exposures, and other design patterns that optimize solar exposures. Although the book is somewhat technical in nature, two chapters, nine and ten, make a valuable contribution to preventive design-oriented strategies.

122. Stone, Laurie. "Solar Cooking in Developing Countries." *Solar Today* 8, 6 (1994): 27-30.

Keywords: renewable, case studies, solar

Many developing countries that used to rely on wood as their main cooking fuel can do so no longer because of drastic changes to the environment. Nine billion acres of what were once forests have now been turned into desert. Many women must rely on wood for cooking, however, and this leads to hardship. Often households spend as much as half their income on fuel for cooking. Some are turning to solar cooking, but there remain barriers to this technology that must be addressed. The price of commercially available cookers from the United States, for example, can be very high, but solar cookers can be made very inexpensively if they are made locally, especially by the people who are going to use them. They make cooking times slightly longer, but do not require the constant attention needed for a wood fire, which must be stoked and generates high temperatures that can burn food. Solar cooking is also much healthier, as smoke from wood fires causes severe eye problems and respiratory diseases. As well as these advantages, the article outlines some problems: solar cooking doesn't work for some traditional foods; and since technological development is often dominated by men, solar cooking is not given priority. In some parts of the world, the fire is also the social gathering place of the community. Stone feels that it is essential for women to become more involved and more educated in these areas because women seem to be more aware of their children's needs and of the problem of malnutrition. Despite the barriers, however, solar cooking is becoming quite successful in many parts of the world.

123. Stone, Laurie. "Building A Dream: Living and Learning Solar 'Off the Grid.'" *Solar Today* 10, 3 (1996): 28-30.

Keywords: renewable, case studies, energy efficient buildings, solar

This is the story of Frank and Patti Phelps, who moved to the Colorado Rockies to retire on a property they had bought. When they discovered that it would cost $60,000 to bring in electric lines to their land, they began to look for alternative energy sources. What they decided on was a home that uses passive solar energy for heat and solar photovoltaics for electricity. The couple, who lived in the city for years, have not had to make any major lifestyle changes, and still use all their electrical appliances. The only other changes they made were installing a gas-powered clothes dryer, refrigerator, and stove. In building their house, they used a quality building envelope with reduced air leakage, superior insulation, and energy-efficient appliances, as well as high quality windows and low-flow shower heads. Their twelve panel, twenty-four volt system gives them a comfortable, although modest lifestyle.

124. Takahashi, Patrick, and Andrew Trenka. *Ocean Thermal Energy Conversion*. Toronto: John Wiley & Sons, 1996.

Keywords: renewable, ocean

The potentially limitless and clean source of energy to be harnessed from the ocean is still, in many cases, under study. This text explores this promising new technology, and although the main focus is on Ocean Thermal Energy Conversion (OTEC) systems, it also looks at tidal, wave, and current energy as well as the exploitation of salinity gradients. OTEC as a concept is described early on in the book. Three different types of systems are looked at (closed-cycle, open-cycle, and hybrid cycle). Challenges involved in the widespread use of this form of renewable energy are addressed, and a brief look is taken at some laws of thermodynamics and fluid mechanics in order to better understand the limitations and potentials of the various technologies. Different forms of ocean en-

ergy technology are examined in greater detail, along with the economics and externalities associated with each. This book is written as a textbook in which each chapter contains a set of aims and objectives at the beginning, as well as some self-assessment questions and answers at the end.

125. Wakhungu, Judi Wangalwa. "Renewable Energy Technologies in Africa: Retrospect
and Prospects." *Bulletin of Science, Technology & Society* 16, 1-2 (1996): 35-40.

Keywords: renewable, case studies, economics, developing countries, policy

This article analyzes several studies related to the implementation of renewable energy
technologies in Africa. They are divided into three groups. The first group is composed
of studies that explore barriers to the dissemination of renewable energy technologies
(RET). These studies tend to overanalyze one particular factor and disregard other important contributing factors. The second group of studies recognizes that barriers to RET
dissemination were caused by several issues but identify only one sector as a solution for
implementation. The third group of studies stresses the importance of an integrated policy approach to enhance the RET dissemination related to rural development in Africa.
Lessons learned from all of these studies can provide valuable guidance for future RET
initiatives in Africa, especially if they are tailored to the specific characteristics of each
region.
    The author notes that it is imperative that RET projects address the particular needs of
users and that user participation in technology design and selection be encouraged. Other
important aspects that need to be addressed are: the creation of a favorable economic
environment; increase in the involvement of the African scientific and technological
community; the adoption of targets for implementing RET; and the development of a
stable institutional structure for coordinating dissemination efforts.

126. Winkler, Mark. "The Real Goods Solar Living Center." *Solar Today* 10, 3 (1996):
20-23.

Keywords: renewable, economy, energy efficient buildings, solar, wind

The Solar Living Center in Hopland, California, began as "a shared vision among the
workers of Real Goods Trading Corporation of Ukiah, California, and company founder
and President, John Schaffer." Sim Van der Ryn of the Ecological Design Institute of
Sausalito, California, was selected to design the building, his associate David Arkin
served as project architect, and Jeff Oldham of Real Goods managed the building of the
project. The building that resulted from this is "a gracefully curving single story structure
that captures the varying hourly and seasonal angles of the sun so effectively that additional heat and light are nearly unnecessary. Innovative landscaping surrounds the passive solar building, which gets all its electricity from photovoltaic and wind power.
Visitors can experience the practicality of numerous applications of solar power, including the generation of electricity and solar water pumping." Further on in the article, Jeff
Oldham looks at the Real Goods Retail Showroom, in terms of solar gain analysis and
control, daylight as form-giver, wind tunnel testing, and economy and ecology.

127. World Energy Council, ed. *New Renewable Energy Resources: A Guide to the Fu-*

ture. London: Kogan Page, 1994.

Keywords: renewable, biomass, geothermal, hydroelectric, ocean, solar, wind

Specialists from many countries, both developed and developing, present their views in this book. Their studies of different forms of renewable energy began in the year 1989 and look at energy consumption for the world in the future, up to the year 2100. Because so many people contributed to this book, and a "bottom up" approach was taken, the result is a very comprehensive, realistic, and global overview of the new forms of energy. The book also tries to take a balanced view of renewable energy in terms of its potential benefits. The "new" renewable energy sources considered are: solar energy, wind energy, geothermal energy, biomass energy, ocean energy, and small hydropower.

128. Wrixon, G. T., A.-M. E. Rooney, and W. Palz. *Renewable Energy—2000*. Berlin: Springer-Verlag, 1993.

Keywords: renewable, biomass, case studies, economics, environmental, photovoltaics, solar, wind

The potential of renewable energy is often debated, since there are differing opinions as to the role it will play in the future. A critical evaluation is done in this work, with a focus on wind energy, solar heating, cooling and daylighting, photovoltaics, and biomass. It also proposes strategies for implementing components and systems to achieve economic operation in different regions of the European Community. The purpose here is to "provide a general review of current technical, economic, market penetration/commercialization status and prospects to the year 2000" for the four renewable energy sources in Europe mentioned above, "and to assess, for each, future R, D & D needs/goals and the potential impact of Community R, D & D programmes to accelerate their technical and economic readiness." For each of the renewable energies looked at, there are several aspects studied closely: a technical and economic review is made, and environmental impacts and public acceptability in the European context are analyzed, as well as commercialization in Europe. The final chapter discusses prospects for the development of renewable energy in Europe to the year 2000.

129. Yaron, Gil, Tani Forbes Irving, and Sven Jansson. *Solar Energy for Rural Communities: The Case of Namibia*. London: Intermediate Technology Publications, 1994.

Keywords: renewable, case studies, economics, photovoltaics, solar

Rural communities often have to make do with insufficient energy simply because they do not have access to it. The problem is that, with areas of low population densities, companies that provide energy services are unable to recover their costs. Also, with traditional forms of energy, such as wood, environmental degradation occurs. Energy experts are therefore looking for new ways to provide these communities with the energy they need. One of these alternatives, solar power, is currently being investigated, primarily in terms of photovoltaic energy, as well as other solar technologies. To achieve a deeper understanding of the relationship between solar energies and rural communities, their impact on rural development, economic viability, consumer willingness to pay, social acceptance, and technical performance of various solar energy alternatives are considered. Although Namibia is the country on which this book focuses, the studies contained here could provide useful information for anyone taking an interest in solar energy for developing countries and rural communities.

130. Zucchet, Michael J. "Renewable Resource Electricity in the Changing Regulatory Environment." *International Renewable Energy* 1 (1995).

Keywords: renewable, electricity, policy

The United States has been a leader in the development of renewable resource electricity since the early 1980s. During the past fifteen years, many renewable technologies have advanced beyond the research stage and into commercial development. However, despite its advances, the commercial renewable energy industry makes up a very small share of the electricity market, and the near-term prospects for more renewable energy development remain uncertain. Much of this uncertainty has arisen in a regulatory environment that is changing to make the electricity industry increasingly competitive. Heightened competition through deregulation and restructuring of electricity generation could inhibit future renewable energy development. New and proposed regulatory policies may also hurt renewables by reducing the importance of their nonmarket benefits in the resource planning process. This article surveys those recent actions and proposals and summarizes their implications for the renewables industry.

Energy End-Use

Demand Side Management

There is considerable overlap between this and the following subsection because most demand side management (DSM) initiatives are aimed at efficiency measures. DSM as a strategy, however, may have many more goals than just efficiency, hence the division into the two subsections. It is probably best if the following section is considered as a whole.

131. Andrews, Clinton J. "An End-Use Approach to Reliability Investment Analysis." *Energy Economics* 14, 4 (October 1992): 248-54.

Keywords: efficiency, end-use

Investments in reliability are increasingly being implemented by users and therefore should be considered by utilities planning strategies to improve electricity reliability. The author presents an end-use approach to the analysis of reliability investments. His approach suggests that, from an economic efficiency perspective, society might prefer a greater emphasis on customer-side reliability investments. Utility end-use data is provided to prove the feasibility of this approach. The paper is divided into the following sections: defining the socially optimal reliability level; incorporating customer investments in reliability; estimating reliability at a point of end-use; the customer's decision-making process; and using end-use data for analyzing reliability investments.

132. Boardman, Brenda. *Fuel Poverty: From Cold Homes to Affordable Warmth.*
London: Belhaven Press, 1991.

Keywords: efficiency, economics, poverty, policy, case studies, fuels

This volume analyzes how poverty affects the quality of home life and access to residential heating. It also provides specific information regarding residential energy efficiency measures. Chapter 1 presents a framework for studying fuel poverty in a historical
context and a profile of the fuel poor. Chapter 4 addresses sources of heat loss in homes,
what added insulation can do, government funded insulation programs, and other similar
initiatives. The subsequent two chapters report on the typical energy efficiency of homes
and the required temperatures within them. It is shown, through a study of household
trends, that the residences of most fuel-poor people do not meet the required temperatures for minimal comfort. Cost-effective energy efficiency improvements and a proposed program for affordable residential heating are proposed.

133. Cavanagh, R. C. "Global Warming and Least-Cost Energy." *Annual Review of Energy* 14 (1989): 353-73.

Keywords: general, case studies, planning, policy

The author notes that, paradoxically, U.S. energy policy is increasing the threat of
global warming. Recent deregulation trends in the energy sector are perceived as minus-
cule steps towards a real solution to the problem of global warming. Based on the experience of particular state utility regulators with the use of "least-cost energy planning," the
article suggests an approach to organize urgent U.S. and international energy policy re-
forms. "Least-cost planning recognizes improvements in the efficiency of energy use as a
major source of additional energy supplies, and seeks fair competition for energy investment dollars between conservation measures and production facilities." The ideas presented in this article provide the basis for the development of a federal least-cost plan that
would act as a catalyst for the implementation of remedial efforts at different government
levels, utilities, and among energy users.

134. Cavanagh, Ralph. "Energy Efficiency Solutions: What Commodity Prices Can't
Deliver." *Annual Review of Energy and the Environment* 20 (1995): 519-24.

Keywords: efficiency, case studies, conservation, environmental

This is a slightly updated version of the author's previous article (see above).

135. Ciccetti, Charles J. "Four Misconceptions About Demand-Side Management." *Annual Review of Energy and the Environment* 20 (1995): 512-18.

Keywords: efficiency, DSM, environment, policy

Although energy efficiency measures translate into significant economic savings,
business and individuals continue to neglect opportunities for investments in this area.
This article examines the opposition of some economists to demand-side management

(DSM) initiatives sponsored by utilities. This opposition is usually based on the following four criticisms: "DSM programs impede market reform and competition; regulators favor conservation even when consumers reject it; the benefits of DSM are grossly exaggerated; and the costs of DSM programs are significantly underestimated." After discussing these criticisms of utility-sponsored DSM programs, the author presents a policy recommendation.

136. Cook, Jeffrey, and Mike McEvoy. "Naturally Ventilated Buildings: Simple Concepts for Large Buildings." *Solar Today* 10, 2 (1996): 22-25.

Keywords: general, case studies, energy efficient buildings, environmental

The authors note that a new design concept has been emerging among architects and engineers for about a decade in England and Scotland. This concept incorporates natural forms of ventilation into the design of large offices and other building types. These designs use passive ventilation based on pressure differentials and the "stack effect," which work "even when the windows are closed." This article analyzes this new type of building design and discusses four case studies: the Inland Revenue Center, the Powergen Atrium, the Queen's Building, and St. John's College Library. Each of these is a variety of naturally ventilated building, and each one is discussed individually in terms of dimensions, architectural design, window strategies, thermal mass, layout, and aesthetic appeal. These buildings are constructed on moderate budgets with low operating costs and demonstrate the view that building occupants should be able to "interact with their immediate work environment, and also should be able to experience the natural environment."

137. de Almeida, Anibal T., et al., eds. *Integrated Electricity Resource Planning*. Boston: Kluwer Academic Publishers, 1994.

Keywords: general, economics, efficiency, environment, case studies, planning, supply and demand

This book contains papers presented at a NATO Advanced Research Workshop on Models for Integrated Electricity Resource Planning. During the workshop, discussions and analysis focussed on cost-effective methodologies to achieve the supply of electricity services at minimum environmental impact and cost. The discussions also centered on new developments in power planning models, which can integrate both supply-side and demand-side actions. Chapter 1 consists of an introduction to integrated resource planning (IRP). Chapter 2 focuses on IRP and the environment, and analyzes quantitative assessments of the environmental impact of different supply-demand strategies. In chapter 3, the central theme is IRP modeling and planning models that deal with uncertainty in the use of multi-criteria approaches. Chapter 4 examines energy-efficient technologies; the authors also analyze several case studies and experiments regarding innovative efficiency concepts undertaken by utilities in several countries. Finally, chapter 5 deals with the potential for electricity savings in the industrial, commercial, and residential sectors from a global perspective.

138. De Neufville, Richard, and Stephen R. Connors. "The Electric Car Unplugged."

Technology Review 99, 1 (1996): 30-36.

Keywords: renewable, economics, transportation, electricity, environmental

This article carefully scrutinizes the electric car. Contrary to popular belief, this new technology is not as economically and environmentally friendly as most people believe. Electric cars are still quite expensive and require subsidies and other forms of financial aid. They also cause pollution, albeit in a different manner than regular cars. Pollution from electric cars originates where the electricity is produced and from their batteries, rather than from fuel combustion within the car. The results obtained from the testing of electric vehicles are not very accurate. They fail to take into account urban driving conditions such as the effect of frequent stopping and starting in traffic and at stoplights and the effects of colder climates and variations in altitude. These factors translate into higher energy requirements that greatly limit the allowable driving distance before the batteries need to be recharged. The authors do not discount the possibility of great improvements in electric car technology in the future, but they maintain that the current state of development of electric cars is not sufficiently advanced to replace gas-powered cars or even to necessitate regulations for their production for commercial use. Other solutions for reducing vehicle emissions should be considered, for example, measures to increase car pooling and the use of buses; promotion of research and development for a broad range of alternative vehicles; and the implementation of programs to replace older and highly polluting vehicles.

139. Kennedy, William, Wayne C. Turner, and Barney L. Capehart. *Guide to Energy Management.* Lilburn, Ga.: Fairmont Press, 1994.

Keywords: general, economics, planning, renewable

The authors define energy management as "the judicious and effective use of energy to maximize profits (minimize costs) and enhance competitive positions." This definition encompasses various measures such as product, equipment, and process design. The objective of the textbook is to demonstrate and teach the basic techniques and tools of energy management. The tools for designing, initiating, and managing energy management programs are put forth in chapter 1. The basics for the study of energy such as terminology, supply and use statistics, and energy accounting are also covered. An overview of the study audit process including suggested forms, equipment, and procedures is provided in chapter 2. Chapter 3, "Understanding Energy Bills," encourages managers to take the time to understand their utility's billing procedures and rate schedules. Techniques and measures of cost-effectiveness are presented in chapter 4 in order to assist with decision-making regarding energy management opportunities. Actual energy management techniques, accompanied by calculations and working examples, make up chapters 5 to 10 for the following systems and processes: lighting; heating; ventilating and air conditioning; combustion processes; industrial wastes; steam generation and distribution; control systems; and computers and maintenance. These chapters are a very good practical resource for step-by-step instructions on how to calculate and design for reduced energy use. Following a discussion on insulation in chapter 11, chapter 12, "Process Energy Management," gives examples of how to evaluate the energy-saving measures that a combination of systems and techniques addressed in the previous chapters. An overview of renewable energy usage and water management concludes the book.

140. Newman, Peter. "Greenhouse, Oil and Cities." *Futures* (May 1991): 335-48.

Keywords: transportation, policy, sustainability, pollution prevention, cities, planning

The author notes that cities with the highest levels of public transportation use also have the lowest levels of carbon dioxide emissions. This conclusion is based on a study of emissions from the use of transport fuels in thirty-one cities in the United States, Europe, Australia, and Asia. The analysis also shows that those cities where automobile use is prevalent have the lowest urban density and the highest level of carbon dioxide emissions. These observations imply that in order to achieve a reduction of carbon dioxide emissions and thereby decrease the risks associated with climate change, significant changes in urban planning must be implemented. The author suggests that these changes should include the use of light rail, traffic calming measures, and the development of urban villages. The article discusses the advantages, experiences, and implications of these practical solutions.

141. Nichols, Albert L. "Demand-Side Management: a Nth-Best Solution?" *Annual Review of Energy and the Environment* 20 (1995): 556-61.

Keywords: efficiency, DSM, economics

Nichols criticizes demand-side management (DSM) programs on the basis that they are economically inefficient. They fail to measure their energy savings appropriately, and analyses used to prove that DSM programs are cost-effective typically omit some potentially important costs. Another criticism is that, even without DSM, utilities already pay their customers to conserve in the form of the price charged for electricity; therefore, end users already have an adequate incentive to choose energy-efficient equipment. Nichols also feels that where the price of electricity is equal to or greater than its marginal cost, DSM rebate programs inevitably involve a significant transfer from rate-payers as a whole to program participants. In addition, the market barriers routinely cited by DSM supporters are far from unique to the electricity market, and virtually none qualify as market failures. Finally, environmental benefits may justify some otherwise cost-ineffective programs, but those benefits should be evaluated explicitly, and policies should be aimed at issues of concern. Emission charges or tradable allowances are a better strategy to internalize environmental effects. However, Nichols suggests that DSM programs can be a useful strategy for developing nations.

142. Nivola, P. S., and R. W. Crandall. *The Extra Mile: Rethinking Energy Policy for Automotive Transportation*. Washington, D.C.: Brookings Institution, 1995.

Keywords: transportation, policy, case studies, pollution prevention, efficiency

The effectiveness of the implementation of the Corporate Average Fuel Economy (CAFE) program in the United States has been a controversial issue. This book argues that a higher federal gasoline tax would be a more effective strategy to reduce fuel consumption and automobile travel than CAFE. The book is divided into two parts: an economic analysis of fiscal instruments versus continued regulation; and an analysis of the politics of both approaches.

The main strengths of this volume are its historical study of CAFE standards and fuel taxation in the United States and its analysis of the politics behind fuel taxation in Great Britain, France, Germany, Japan, and Canada. The authors conclude that although a substantial increase in the Federal Gasoline Tax is not a feasible option in the near future, it nevertheless constitutes a desirable policy choice. They believe that this conclusion will become more accepted as fiscal deficits persist and as policy-makers try to promote new automotive technologies in the face of declining oil prices.

143. Price, Trevor, and Doug Probert. "Environmental Impacts of Air Traffic." *Applied Energy* 50 (1995): 133-62.

Keywords: environmental, case studies, policy, pollution prevention

The authors state that before the 1990s environmental degradation caused by emissions from aircraft use was considered to be negligible. The article presents some of the most well-known consequences of emissions from aircraft. It also analyzes current UK environmental legislation regarding aviation and developments in aircraft technology. Finally, the authors provide a set of recommendations to improve the sustainability of air-transportation policy in the UK and elsewhere.

144. Probert, S. D. "Environmentally and Energy Responsible Universities?" *Applied Energy* 50 (1995): 69-83.

Keywords: efficiency, environmental, policy, case studies, sustainability

This article discusses the implementation of integrated environmental, health, and energy strategies in universities. Although the focus is on British universities, most of the analysis and several of the suggestions presented by the author could be applied to universities in many other countries. The article is divided into the following sections: external influences, educational responsibility, moral and financial incentives, considerations behind an appropriate mission statement, a typical commitment manifesto, the degree of implementation, and how to sustain an environmental management campaign.

145. Reddy, B. Sukahara. "Economic Evaluation of Demand-Side Management Using Utility Avoided Costs." *Energy* 21, 6 (June 1996): 473-82.

Keywords: economics, conservation, DSM

The author believes that demand side management (DSM) has become an important policy issue and that its implementation can promote efficient electricity use. However, the use of DSM in developing nations, such as India, is not yet very common. In order to encourage the use of DSM methods in those nations, it is necessary to provide case studies that compare the benefits and costs of DSM options to the more conventional alternative of increasing generating capacity. This article evaluates the avoided marginal costs of power generation related to various DSM options for the industrial sector of Maharashtra, India. It shows that through the implementation of DSM options, 1570 MW of peak demand and 12510 GW of energy overall can be saved by the year 2000. These energy savings translate directly into economic savings for the utility and the consumers. The value of this case study is that it clearly illustrates that there are viable alternatives to the conventional view that utilities can only profit if they increase their electricity sales.

146. Riley, Robert Q. *Alternative Cars in the 21st Century: A New Personal Transportation Paradigm.* Warrendale, Pa.: Society of Automotive Engineers, 1994.

Keywords: transportation, efficiency, electricity, environmental, fuels

This book addresses the issue of transportation, specifically, the burden that cars place on the environment. In view of attempting to reduce our use of energy and resources related to the growing automobile industry, various types of alternative cars are presented and analyzed. Chapter 1 looks at the automobile today, its place in our society, its impact on the environment, its comparative use in various countries, and ways to reduce this impact by conventional means. In chapter 2, alternative transportation for the twenty-first century is discussed. Various types and models are presented, such as small cars, low-mass cars, fuel-efficient vehicles, ultralight automobiles, the passenger car, the commuter car, the narrow-lane vehicle, the urban car, and the sub-car. These are looked at in terms of emissions, design, layout, and handling. Chapter 3 examines fuel economy, aerodynamics, and the fuel-efficient powertrain, as well as various aspects of the engine. In chapter 4, alternative fuels are investigated, such as alcohol fuels, hydrogen, electricity, and natural gas, in terms of their efficiency, performance, and compatibility with current motor design. Chapter 5 focuses on electric and hybrid vehicles of different types and configurations, while chapter 6 takes a look at three-wheeled cars. In chapter 7, the issue of safety and low-mass vehicles is examined through accident statistics, safety engineering, crash dynamics, and crash protection. The final chapter looks at alternative cars in Europe, from minicars and the European transit systems, to the future of urban cars and commuter cars in Europe.

147. Schaeffer, Loretta. "Support for Alternative Energy in Asia." *Resources, Conservation and Recycling* 12 (1994): 57-62.

Keywords: DSM, case studies, environmental, renewable, solar

This article represents a useful summary of demand side management (DSM) initiatives and renewable energy programs that are funded by the Asia Alternative Energy Unit (ASTAE). The first section provides background information on the energy situation in Asia. It is stated that DSM and renewable energy technologies can deal with Asia's rapidly growing energy demand while reducing the harm inflicted on the biosphere. The article goes on to sketch the organization and objectives of the World Bank's ASTAE. DSM initiatives in Thailand, the Philippines, Indonesia, and Laos are very briefly outlined. Similarly, the ASTAE's renewable energy work program's activities in India and Indonesia are summarized. The lessons learned from the experience of various solar photovoltaic rural electrification programs are outlined. Barriers to DSM and renewable energy development and the ASTAE's role in overcoming them are also discussed.

148. Stepanov, V. S. *Analysis of Energy Efficiency of Industrial Processes.* Berlin: Springer-Verlag, 1993.

Keywords: efficiency, economics, energy analysis

This book examines methods for studying energy use efficiency in technological processes and estimates the theoretical and actual energy reserves in a given process, technology, or industrial sector, on the basis of their complete energy balances. It also presents the results of long-term studies in the field, conducted in ferrous and non-ferrous metallurgical plants using exergy analysis. The concepts and laws of thermodynamics that form the basis for the exergy method, and techniques for drawing and analyzing the complete energy balance of a process, are outlined. Methods for computing the chemical energy and exergy of substances, developed with the participation of the author and differing from those proposed by other researchers, are also provided. The methodology used in the complex analysis of processes and plants (including thermodynamic and techno-economic analyses) is examined, using thermal secondary energy sources as an example. Examples are provided of how to draw up the complete energy balance of individual technological processes and complexes of ferrous and non-ferrous metallurgy, determine their efficiency, analyze their losses, and identify irrevocable losses as well as losses that are theoretically recoverable. An estimate is given of the energy reserves of various industries (comprised of technological processes) using standard methodological principles based on the concept that energy conservation in an industry results from more efficient energy use in its individual technological processes.

149. Van Engelenburg, B. C. W., T. F. M. Van Rossum, K. Blok, and K. Vringer. "Calculating the Energy Requirements of Household Purchases." *Energy Policy* 22, 8 (1994): 648-56.

Keywords: efficiency, case studies, consumption, economics

A new and more practical means of calculating the amount of energy required for various products and processes or "consumption items" is discussed in this article. The method used combines the best characteristics of two existing methods used to determine the energy requirements for goods and services. These are: "the physical approach provided by process analysis, and the economic-statistical approach provided by input-output analysis." A description of the two approaches follows, and the process used in order to combine them is laid out in ten steps. This process is shown through the use of a good example involving the production and use of a domestic refrigerator.

150. Wene, C. O. "Energy-Economy Analysis: Linking the Macroeconomic and Systems Engineering Approaches." *Energy* 21, 9 (1996): 809-24.

Keywords: economics, energy analysis

Methods are being sought for the joint analysis of engineering and the economy. In this paper, the informal linking or softlinking of macroeconomic and systems engineering models that can enable such studies to be carried out are discussed. It is argued that formalized language describing the area of overlap between the two models is a necessary aspect of such linking. The characteristics of this language are outlined in this article, and demonstrated as they pertain to the softlinking of a macroeconomic model (ETA-MACRO) and a systems engineering model (MESSAGE).

151. Wernick, Iddo K., and Jesse H. Ausubel. "National Materials Flows and the Environment." *Annual Review of Energy and the Environment* 20 (1995): 663-92.

Keywords: environmental, consumption

The circulation of large amounts of material goods is required for the functioning of modern society and to satisfy the growing population's wants and needs. It is estimated that for Americans, this figure is 50 kg per person per day. Focusing on the U.S. economy, this article develops and tests a comprehensive framework to order the materials flows related to demand for energy, construction, and food. It assesses and quantifies inputs to the national economy, outputs, foreign trade, and wastes from resource extraction, using mass measures of these flow components. Atmospheric emissions and materials embedded in long-lived structures dominate outputs, with smaller contributions from solid wastes and dissipated materials. The authors conclude that an effective way to assess environmental performance at the national level is through the "consistent, periodic accounting of physical flows. Improvements in the collection and organization of the data supporting national material accounts will further their utility."

152. Williams, Robert H., and Eric D. Larson. "Materials, Affluence, and Industrial Energy Use." *Annual Review of Energy* 12 (1987): 99-144.

Keywords: economics, case studies, consumption

This article examines the role of materials in the economy and the many debates over the future of these materials. The question addressed here is whether demand will grow at all, remain stagnant, or go into a long-term decline. Williams and Larson suggest that a "structural shift" has occurred that is an indication of a new trend. If such a shift to the production of increasingly refined and complex goods has indeed taken place, the energy needs of industry will be lower than was forecast. The article looks into "recent analyses of ongoing trends away from the materials-intensive industries, and presents the authors' own analysis of the trends and the future outlook for basic materials, and an assessment of the implications for future industrial energy use in the United States."

153. Wirl, Franz. "On the Unprofitability of Utility Demand-Side Conservation Programmes." *Energy Economics* 16, 1 (January 1994): 46-53.

Keywords: efficiency, conservation, DSM, economics, planning

Demand-side conservation programs are the subject of this paper, particularly with regard to their economics. Several of the proposals that have been made suggest that conservation should be an integral part of the least-cost planning process. This is reinforced even more strongly as a result of market failures and high production costs. The weakness in this argument is the assumption that when utilities make moves to promote conservation, the public, and certainly consumers, will not react any differently. This problem is addressed in the paper and solutions presented. Such interactions are considered in a case study of a public utility and its consumers. The case study demonstrates that "strategic interactions render unprofitable demand-side conservation programmes that appear profitable by conventional criteria." The explanation for this is that "rational consumers will respond strategically to the prevalence of utility sponsored conservation programmes by minimizing their own expenses for energy efficiency."

End-Use Efficiency

154. Barker, Brent. "Energy Efficiency: Probing the Limits, Expanding the Options." *EPRI Journal* 17, 2 (1992): 14-21.

Keywords: efficiency, end-use, policy

   This article aims to determine the place that energy efficiency should occupy in future energy planning. The author summarizes the views expressed by several energy specialists at a seminar sponsored by EPRI. He states that energy efficiency was considered as having very limited potential in the 1970s. Despite that view, energy efficiency has become an influential aspect in every end-use sector and an essential part of energy planning. He suggests some incentives that would increase the attractiveness of energy efficiency measures and highlights their value to society. Also examined are the great position of influence that energy efficiency could have on growing environmental concerns and who should pay for its implementation costs.

155. Bullard, Clark W. "Energy Conservation in New Buildings: Opportunities, Experience, and Options." *Annual Review of Energy* 6 (1981): 199-232.

Keywords: supply and demand

   The author notes that more than one third of U.S. energy is consumed in buildings, where it is used mostly for space conditioning and water heating. This energy consumption will increase if new construction projects are not well designed to use energy more efficiently. Estimates by the U.S. Department of Energy "indicate that the rate of new nonresidential construction will increase from 1.4 billion square feet annually to a rate of about 2.0 billion in 2000. The rate of new residential construction is expected to decline from recent rates of about 2.3 million units to about 2.0 million in 2000." This article examines the demand side of energy use with a focus on construction. Topics include: opportunities for conservation in new buildings; the regulatory approach and the ECPA; and conservation policy options.

156. Carmody, John, Stephen Selkowitz, and Lisa Heschong. *Residential Windows: A Guide to New Technologies and Energy Performance*. New York: Norton, 1996.

Keywords: efficiency, energy-efficient buildings

   Windows can play quite an important part in energy conservation and are therefore very relevant for sustainability efforts. This book is about the residential use of windows; its purpose is to assist consumers, designers, and builders to understand new window products and their performance implications. The authors introduce various window technologies and analyze the implications of these new technologies on residential design. A guide for selecting appropriate windows is provided as well. The first chapter provides an overview of the volume. Chapter 2 presents the basic energy performance characteristics of windows and describes how these are determined. Chapter 3 describes glazing materials and new technologies in detail. Chapter 4 addresses the complete cycle of window assembly and includes details about window installation and frame materials. Chapter 5 examines traditional window design issues and new design implications related to the use of high-performance windows. In chapter 6, the entire range of window selection considerations are summarized, and a checklist is provided to be used by designers, builders, and homeowners. The authors also describe methods for assessing annual energy performance.

157. Economic Commission for Europe. *Energy Efficient Design: A Guide to Energy Efficiency and Solar Applications in Building Design.* New York: United Nations, 1991.

Keywords: renewable, case studies, efficiency, energy-efficient buildings, solar

This is a guide to the principles behind energy efficient buildings. The authors maintain that the design of buildings is based on the following five steps: understanding the need; defining the brief; design activity; implementing the design; and monitoring buildings in use. The first two of these stages are outlined in this book and their analysis helps to "understand the need for energy efficiency and to define a brief which keeps energy efficiency in mind by explaining the basic characteristics and interactions of the different factors involved." The fifth stage, monitoring buildings in use, outlines various methods available to monitor buildings. The emphasis of the book is on principles and not on details of design, and therefore instructions for designing an energy-efficient building, or descriptions of what one should look like, are not included. The first section of this volume examines energy efficiency in terms of strategies, human factors, environment, and economics. Section 2 discusses architecture, building elements, and solar applications. Section 3 examines monitoring systems, with examples from several countries. The concluding section deals with computer software for energy-efficient design.

158. Grady, Wayne. *Green Home: Planning and Building the Environmentally Advanced House.* Camden East, Ontario: Camden House Publishing, 1993.

Keywords: efficiency, case studies, energy-efficient buildings

This book describes the entire process involved in the building of the "Greenhome," a house built in Waterloo, Ontario. The house is part of the Advanced Houses program organized by Energy, Mines, and Resources Canada. The Greenhome incorporates the most advanced technologies in energy efficiency and ideas related to building in an environmentally sound manner. A special group of environmentally conscious engineers and builders collaborated to choose all of Greenhome's materials and systems. Their main goal was to ensure that it "will have less impact on the land than any other house ever built in North America." The author followed and recorded every step related to the building of the Greenhome. Every aspect of the house was built with energy efficiency and cost in mind, and with respect for the environment as the main guiding principle. The book is divided into the stages in the building of the house: survey, ground work, foundation, frame work, mechanicals, finishings, and opening. The volume also examines some of the most efficient heating systems, windows, and appliances.

159. Hollander, Jack M., and Thomas R. Schneider. "Energy-Efficiency: Issue for the Decade." *Energy* 21, 4 (1996): 273-87.

Keywords: efficiency, economics, policy

The authors note that during the past twenty years the implementation of energy-saving measures has profited the economy and helped to protect the environment. However, the implementation rate of these measures is decreasing while controversy about their validity persists. This article analyzes the nature of the controversy behind energy-efficiency initiatives, and the potential role of "market forces, government intervention, and technological innovation in determining future progress." The article also explores the future of energy-efficiency programs within the context of opposing forces such as current deregulation trends, growing concern about environmental issues, and falling energy prices. The authors highlight the role of R&D investment as a "key ingredient in the development of energy-efficient technologies" and note that it is imperative to preserve a strong partnership between the private sector and governments to ensure future progress in the field. The article is divided into the following sections: the history of energy-efficiency; energy efficiency today; the future of energy efficiency; technologies, mandatory standards, and the market; energy efficiency and the environment; and developing a long-term solution.

160. Holm, Dieter. *Energy Conservation in Hot Climates*. London: Architectural Press, 1983.

Keywords: efficiency, alternative fuels, conservation, energy efficient buildings

This book analyzes, in a pragmatic manner, issues related to energy conservation in hot climates. It will be of particular interest to readers concerned with energy use in buildings. The author illustrates several ways to reduce energy demand and losses, and emphasizes the adequate use of passive and active energy systems by focusing on aspects such as lifestyle, available technology, and costs. One of the goals of this volume is to suggest alternatives that are realistic and practical. Chapter 7 analyzes how methane can be produced, stored, and used to replace fossil fuels.

161. International Energy Agency. *Energy Efficiency and the Environment*. Paris: OECD, 1991.

Keywords: efficiency, economics, end-use, environmental, policy

This book addresses the issue of energy efficiency and looks at ways in which we can improve this field further in terms of the environment and the economy. Various factors that influence the cost-effectiveness of energy efficiency measures, such as technical innovations and market forces, are analyzed. Measures implemented to enhance the widespread adoption of energy efficient technologies and the role of energy-specific information on efficiency-related investment are evaluated. The book is divided into four sections that examine the following issues: energy end-uses and carbon dioxide emissions; energy demand and efficiency trends; potential energy efficiency solutions; and policy instruments for improving energy efficiency. The authors note that the most significant effect of the implementation of energy efficiency measures would be a reduction in the production of carbon dioxide. The following policy instruments for improving energy efficiency are discussed in detail: information, regulation, price setting, and taxation.

162. International Energy Agency. *Industrial Energy Efficiency: Policies and Programmes*. Washington, D.C.: U.S. Department of Energy, 1994.

Keywords: efficiency, DSM, economics, environmental, policy

The International Workshop on Industrial Energy Efficiency served as the foundation for this book. One hundred and eighty people from government, industry, utilities, universities, and other organizations came together to present papers on opportunities for improving energy efficiency and to examine some successful industrial energy efficiency approaches. The workshop encouraged discussions between government and industry representatives. One of the main goals was for government and industry to "work together to meet common energy, economic and environmental goals." The following areas of government policy related to industrial energy efficiency were covered: information and technical assistance programs; technology development and commercialization strategies; recognition and commitment activities; industrial demand-side management programs; tax and fiscal approaches; and efficiency standards and regulatory policies.

163. Joskow, Paul L. "Utility-Subsidized Energy-Efficiency Programs." *Annual Review of Energy and the Environment* 20 (1995): 526-34.

Keywords: efficiency, economics, end-use

This article begins by describing how problems in the electricity sector of developing countries are different from those of developed countries. As an example, it is noted that electrical utilities in the United States tend to be economically healthier because they operate in a mature market that is relatively predictable. Also explained are some of the reasons behind utility-subsidized energy-efficiency programs. Four arguments related to these programs are mentioned by the author: the existence of large "untapped" economic opportunities to conserve energy; the reality of numerous market barriers and market imperfections that prevent consumers from taking advantage of these energy-efficiency opportunities; the fact that utilities are in a unique position to overcome these market barriers and imperfections; and the problems arising form the utilities' failure to invest in end-use energy efficiency. The last section examines the fairly uniform response to these four arguments by economists.

164. Kozloff, Keith Lee, and Roger C. Dower. *A New Power Base: Renewable Energy Policies for the Nineties and Beyond*. Washington, D.C.: World Resources Institute, 1993.

Keywords: renewable, case studies, economics, environmental, policy

This report examines current renewable energy policies and identifies strategies to enhance environmental and economic benefits from renewable energy sources in the United States. The overall strategy is to identify market and institutional barriers that have enhanced or impeded the development of renewable energy in the United States, in order to improve the effectiveness of future policies. Section 1 looks at creating a renewable energy future—specifically, what opportunities and constraints renewables face, what policy experts teach, and how to enhance the benefits of renewable energy. Section 2, "Recognizing Renewable Energy's Benefits," examines environmental protection, economic sustainability, increased energy security, an equitable distribution of economic benefits, and banking on renewables in uncertain times. Section 3, "Reforming Energy-Price Signals," discusses internalizing social costs and reforming energy subsidies. Section 4, "Revamping Utility Decision Making," analyzes major federal and state initiatives in electricity-supply regulation, the strengths and weaknesses of renewable energy, alternative approaches to influence utility decisions, and siting characteristics. Section 5, "Changing Energy Users' Investment Incentives," discusses how to select the right tools to promote renewable energy investments, how to improve policy implementation, and how to determine appropriate institutions to promote demand-side applications. Section 6, "Accelerating Investment in Renewable Energy Commercialization," examines the following topics: reasons that justify government involvement; improving public investments; attracting private investments in commercialization; and developing coordinated commercialization plans.

165. Landsberg, Dennis R., and Ronald Stewart. *Improving Energy Efficiency in Buildings: A Management Guide.* Albany: State University of New York Press, 1980.

Keywords: efficiency, conservation, energy analysis, energy efficient buildings

"The purpose of this handbook is to present a systematic approach to reducing energy consumption in buildings. There are four parts to this process. First, patterns of energy usage are identified by means of energy audit techniques. Second, alternatives which would reduce energy consumption in each building system are examined. These may be simple schedule changes or complex building retrofits. Third, modifications to the building structure systems or schedules are rated according to priority in order to keep alterations within budgetary limits. Fourth, energy conservation is made a permanent part of building operations." The handbook is broken down into five sections. The first contains discussions of the energy crisis, the future availability and cost of energy, preliminary building energy analysis, and approaches to saving energy. The second part contains a description of systems and factors that contribute to energy consumption. In section 3, instructions for performing advanced energy audits are provided, including sample worksheets, suggestions for measurement equipment, time estimates, and skills requirements. Section 4 contains a discussion of building system modifications and is divided into two phases. Phase 1 presents guidelines for modifications which can be evaluated by simple calculations. Phase 2 describes the evaluation of complex energy-saving options through computer simulations. The merits and applications of each approach are also discussed. The five main sections are followed by six appendices which provide background information for use in applying the techniques outlined in the handbook.

166. Meckler, Milton, ed. *Retrofitting of Buildings for Energy Conservation*. 2nd ed. Lilburn, Ga.: Fairmont Press, 1994.

Keywords: cogeneration, conservation, economics, planning

This comprehensive reference text places special emphasis on hands-on and operating experience in surveying, auditing, planning, design, construction, and maintenance of energy-conserving equipment, systems, and processes that have been employed and retrofitted within existing operations. It is centered around practical measures that can be easily implemented, as well as innovative techniques and challenging new approaches put together by a group of experts whose aim was to help reduce energy consumption. Section 1 of the book covers such topics as utility rates and retrofit strategies, managing the energy use survey and audit, economic analysis of energy retrofit projects, and reconciling actual retrofit performance with projections. Section 2 covers design and installation strategies in terms of HVAC system analysis, centralized vs. distribution fan systems in high-rise buildings, cogeneration as a retrofit strategy, establishing variable-air-volume airflow rates on retrofit projects, estimating demand-control savings on retrofit projects, and commissioning for retrofit projects. Section 3 delves into energy audits and management, distributed direct digital control energy management for the 1990s, demand-side management, the energy services industry, and a comprehensive management approach to lighting system upgrades. Section 4 analyzes case studies of retrofit projects for energy conservation. Included are: institutional rehabilitation for energy savings; an energy audit program for demonstrating schoolhouse energy efficiency; a commercial retrofit project in East Texas State University; a heat reclamation system project for a chemical laboratory; industrial energy management; industrial energy observations and opportunities; industrial combustion retrofit opportunities; and industrial retrofit economics.

167. Mills, Evan, and Art Rosenfeld. "Consumer Non-Energy Benefits as a Motivation for Making Energy-Efficiency Improvements." *Energy* 21, 7 & 8 (July/August 1996): 707-20.

Keywords: efficiency, economics

Other than the provision of employment and the generation of energy, few other benefits are provided by electric power plants, coal mines, oil pipelines, or other energy supply-systems. The authors remark that, in stark contrast, technologies that enhance energy efficiency provide various non-energy related benefits. These technologies can increase a nation's economic competitiveness, create new jobs, and improve environmental quality. From the perspective of energy consumers, non-energy related benefits influence the adoption of alternatives that increase energy efficiency. Specific illustrative examples mentioned in this article are: high efficiency windows, energy-efficient lighting, space conditioning, ventilation, and indoor air quality.

168. National Audubon Society. *Audubon House: Building the Environmentally Responsible, Energy-Efficient Office*. New York: John Wiley & Sons, 1994.

Keywords: efficiency, energy-efficient buildings, sustainability

This book follows the efforts of several architects who worked together to create an environmentally responsible office. It is divided into two parts: "Toward a Sustainable Architecture," and "Inside Audubon House." The book examines all aspects related to the building such as lighting, cooling, ventilation, materials, and recycling. All these elements contribute to create a healthy, pleasant, and environmentally sound workplace. This is a practical case study for readers interested in the implementation of energy efficiency measures.

169. Ontario Ministry of Energy. *The Advanced House*. Toronto: Government of Canada, 1990.

Keywords: efficiency, case studies, energy-efficient buildings

This report documents the building of the "Advanced House" in Brampton, Ontario. The purpose for the construction of this house was to demonstrate new products and technologies that reduce energy use. The house consumes only slightly more than one quarter of the energy used by a regular home. Some of the products employed in its construction were so new that they were not yet commercially available, although many others could be bought at common hardware stores. The "Advanced House" uses the following energy-saving technologies: high-performance energy-efficient windows; an integrated mechanical system that replaces a regular furnace, hot water tank, air-conditioner, and ventilation system; energy-efficient appliances, which use 20 to 60 percent less energy than regular appliances; high levels of insulation; airtight construction; a two-story passive-solar sun space; energy-efficient fireplace; and an energy-monitoring system. The report examines all of these features in ample detail.

170. Organization for Economic Co-Operation and Development. *Energy Conservation in IEA Countries*. Paris: OECD/IEA, 1987.

Keywords: efficiency, case studies, conservation, policy

As the price of oil rose in the 1970s, so did the need for energy conservation measures. Since that time, energy conservation has become an important issue in government energy planning. This volume examines how energy conservation has affected the present energy situation, available means for government to promote energy efficiency, the future potential for energy efficiency, and factors that may hinder their implementation. Part A presents an overview of the study; part B examines strategies to improve energy efficiency, energy trends, conservation opportunities and obstacles. Part C examines conservation policies and their effectiveness. The study concludes that carefully planned government policies can help to reduce market imperfections and regulatory obstacles that impede the development and implementation of cost-effective measures to conserve energy.

171. Ricketts, Jana, ed. *Energy, Business, and Technology Source Book*. Lilburn, Ga.: Fairmont Press, 1997.

Keywords: efficiency, energy-efficient buildings, economics, end-use, case studies

This volume contains papers by eighty-seven different authors covering business and technological issues related to energy-efficient technologies and operational improvements for buildings and plants. It includes the latest strategies in energy management and emerging technologies for energy generation illustrated by numerous case studies. It provides ample guidance on how to take advantage of performance-based contracts and innovative financing within the present context of utility restructuring.

172. Roaf, Susan, and Mary Hancock, eds. *Energy Efficient Building: A Design Guide.* Cambridge, Mass.: Blackwell, 1992.

Keywords: efficiency, case studies, energy-efficient buildings

This book is based on "The Oxford Energy Lectures 1990-1991" and contains papers by specialists in the areas of physics, chemistry, energy, and design. Its first section examines design standards for the indoor environment, thermal comfort in terms of energy conservation, the control of health and comfort in the built environment, and daylight and energy. The second section covers environmental control in energy-efficient buildings, various building materials for health and the environment, and efficient and effective lighting. Part 3 examines the building envelope: material costs for low-energy buildings, the effects of smart glazing on design and energy, and a case study of atria. Part 4 addresses domestic insulation, while part 5 concentrates on energy-efficient housing and design. Part 6 explores solar design in non-domestic buildings, including low energy strategies, and includes a case study for innovative design of a working environment.

173. Vine, Edward, and Drury Crawley, eds. *State of the Art of Energy Efficiency: Future Directions.* Washington, D.C.: American Council for an Energy-Efficient Economy, 1991.

Keywords: efficiency, energy-efficient buildings, environmental, planning

The papers that make up this book are later versions of those presented at the American Council for an Energy-Efficient Economy (ACEEE) 1990 Summer Study on Energy Efficiency in Buildings. The objective here is to make people more aware of the latest research on energy efficiency in buildings. Issues such as the design and implementation of government and utility programs, appliance standards, the analysis of energy for buildings, and resource planning are considered. The following topics are discussed in some detail: integrated resource planning for electric and gas utilities; environmental externality costs in electric utility resource planning and regulation; utility conservation and load management programs; U.S. energy efficiency standards for residential appliances; the impact of conservation on household energy demand; end-use load shape data: collection, estimation, and application; and building energy performance monitoring.

174. Zackrison, Harry B. *Energy Conservation Techniques for Engineers.* New York: Van Nostrand Reinhold, 1984.

Keywords: efficiency, alternative fuels, conservation, economics

Written as a practical guide, this book could prove useful for engineers as well as others interested in conservation techniques. Various techniques and systems for saving energy and costs are examined. The first chapter of the book addresses ways in which innovative electrical design can save energy. Chapter 2 seeks to maximize energy conservation by applying "interdisciplinary engineering" and "architectural value engineering analysis." Furnaces, motor control centers, power sources, emergency power, HID lighting, etc. are looked at. Chapter 3 includes the demand control of building electrical HVAC systems. Chapter 4 looks at energy and cost savings of different colored roofs (light vs. dark), exhaust fans, and shading. Chapter 5 discusses water conservation techniques that save both energy and money. Chapters 6 through 8 look extensively at lighting technology and various types of lighting from the past and present, and possibilities for the future. Chapter 9 also discusses lighting, concentrating on methods that combine energy conservation with aesthetics. Chapter 10 looks at energy conservation in a computer facility and a major production system's lighting systems. Chapter 11 discusses alternate energy fuels. Chapter 12 takes a humorous look at the future office of an engineering and architectural firm.

Energy Policy for Sustainability

175. Abdalla, Kathleen L. "Energy Policies for Sustainable Development in Developing Countries." *Energy Policy* (January 1994): 29-36.

Keywords: efficiency, policy, sustainability

The predominant forms of energy use in developing countries are directly responsible for environmental degradation, and complicate the implementation of strategies for sustainable development. This article looks at various short- and long-term policies that may help to develop a more sustainable energy path. Some of the short-term policies discussed are: promotion of energy efficiency, tradable permits, tax incentives, budget controls, and direct legislation. Long-term policies related to rural and social development, transportation, and energy supply are also analyzed within the context of development issues. Some of the economic and institutional barriers to the implementation of all these policies are also examined.

176. Adams, D. K., ed. *Facing the Energy Challenge: Perspectives in Canada and the United Kingdom*. Keele: Keele University Press, 1993.

Keywords: alternative energy, case studies, efficiency, policy, supply and demand

This comprehensive study came about as a result of the Canada-UK Colloquium of 1992, which included parliamentarians, officials, business people, financial, industrialists, media experts, and academics. Some of the issues discussed included issues of supply and demand in the energy industries; alternative energy sources; the role of government policies and regulation in energy strategies; the financial conditions of energy industries; and strategies for the future. In addition, several energy issues were analyzed in terms of politics and economics. The last paper deals with energy efficiency research and development in the United Kingdom.

177. African Energy Policy Research Network (AFREPREN). *African Energy: Issues in*

Planning and Practice. Atlantic Highlands, N. J.: Zed Books, 1990.

Keywords: case studies, conservation, developing countries, fuels, planning, policy, renewable

This volume contains summarized versions of original papers presented at the AFREPREN workshop that took place in Gaborone in 1989. These papers were written by energy specialists and government policy-makers. They focus on six areas of importance to Africa: renewable energy technologies, bioenergy, electricity, coal and gasification, oil and natural gas, institutions and planning. The renewable energy technologies section covers renewable energy technologies in several African regions such as Uganda, sub-Saharan Africa, and Bujumbura, Burundi. This section also analyzes policy and research on rural energy technology. The bioenergy section looks at the rural and urban energy situations in the central region of Sudan and Rwanda, as well as fuelwood surveys in Botswana. The electricity section discusses case studies on energy demand, consumption and conservation, hydro and grid extension alternatives for rural electrification, and hydro electricity and peat as fuel alternatives. The coal and gasification section looks at options for power generation, issues and prospects for coal use in rural households, and coal availability and examines the substitution of coal for woodfuel and charcoal. The oil and natural gas section discusses low grade ethanol as a kerosene substitute for cooking and the acquisition of technical skills in the Tanzanian oil sector. The institutions and planning section analyzes foreign assistance for skill development in energy projects; the role of energy models as policy tools; energy demand management and an energy demand forecast for Ethiopia.

178. Anderson, Dennis, et al. "Roundtable on Energy Efficiency and the Economists: Six Perspectives and an Assessment." *Annual Review of Energy and the Environment* 20 (1995): 493-573.

Keywords: efficiency, economics, electricity, DSM, policy

This group of papers provides the reader with six different perspectives on energy efficiency in relation to the electricity industry, offering a broad perspective on how energy efficiency policies can be improved. The papers cover the following topics: the possibility of energy-efficient strategies emerging from market forces alone; the economic ramifications of utility-subsidized energy-efficient programs; and market failures associated with energy efficiency programs. A final paper summarizes the nexus between the market economy and energy production and use.

179. Anderson, Dennis. "Energy Efficiency and the Economists: The Case for a Policy Based on Economic Principles." *Annual Review of Energy and the Environment* 20 (1995): 495-511.

Keywords: economics, efficiency, policy

According to this paper, the key to reconciling high levels of energy consumption with greatly reduced pollution lies in achieving a policy based on economic principles. This policy must contain the following eight elements: "(1) a stable and growing economy; (2)

price efficiency; (3) corporatization and independent regulation of the energy industry; (4) openness to private investment; (5) demand management programs emphasizing customer information services and energy services willing to work with utilities, manufacturers of electrical equipment and machines, and with designers of buildings; (6) environmental taxes and regulations to reduce external costs; (7) an effective approach to rural energy supply programs in developing countries; and (8) research and development into new energy production and end-use technologies." These eight policies are then analyzed in detail.

180. Bellarmine, G. Thomas. "Energy Management Techniques to Meet Power Shortage Problems in India." *Energy Conversion and Management* 37, 3 (1996): 319-28.

Keywords: case studies, conservation, alternative energy

Despite generous investments in the power sector, India suffers from energy shortages. This article looks at different energy management techniques to alleviate this problem. Several suggestions are made, such as the increased importance of energy conservation and the notion that energy saved is equivalent to energy produced. As conventional energy sources become less desirable, other sources such as solar, wind, bio-mass, ocean thermal, tidal, geothermal, and others must be developed further. Common-sense management techniques are suggested: by modifying working hours for lift irrigation pump sets, for example, demand conservation techniques can be applied. Work schedules should be changed to coincide with periods of low energy needs for industries, and energy audits should be used to decrease the industry's demand and consumption of energy. The article claims that if these suggestions are followed power shortages could be avoided.

181. Dandridge, Cyane B., Jacques Roturier, and Leslie K. Norford. "Energy Policies for Energy Efficiency in Office Equipment." *Energy Policy* 22, 9 (1994): 735-47.

Keywords: efficiency, case studies, consumption, policy

Energy efficiency has become a popular means of reducing energy consumption around the globe. This study looks into how energy can be reduced in the use of office equipment through energy policies and programs. It provides case studies from several European countries: Switzerland, Sweden, Denmark, UK, France, the Netherlands, the European Community, and Norway, as well as Japan and the United States. Each of these is examined in terms of federal government programs, energy consumption, test procedures, industry protocols, and research efforts. There are two sets of proposed procedures for testing energy consumption of copiers, fax machines, and printers. The results of this paper may help to guide future work and research on this subject, as well as suggest areas for cooperation.

182. Daniel, Ronald J., ed. *Ontario Hydro at the Millennium: Has Monopoly's Moment Passed?* Montreal: McGill-Queen's University Press, 1996.

Keywords: efficiency, economics, nuclear, electricity, case studies, policy, energy analysis

The restructuring of Ontario Hydro, Canada's largest public utility, is a complicated policy challenge that is subject to many diverging views and interpretations. Policy analysts tend to agree that some changes in the form of industry de-integration and privatization are needed to improve Ontario's electricity supply. However, there is less consensus on the precise measures that should be implemented to achieve a more sustainable and efficient system. Controversy exists on issues such as the choice of competition strategies; the proportion of assets that need to be privatized; the fate of Ontario Hydro's nuclear industry; the quantification of stranded costs from nuclear investments and ways to deal with these costs. This volume presents ten papers prepared by legal experts, policy-makers, economists, and stakeholders that deal with these and other issues. The goal of these works is to clarify the implications of different strategies to restructure Ontario Hydro.

183. De Baun, R., Nikos Frangakis, and A. D. Papayannides, eds. *Energy Options in a Changing World: A European Perspective.* Cambridge, Mass.: Kluwer, 1995.

Keywords: case studies, economics, fuels, policy

The Athens Conference, "Energy Options in a Changing World," was organized to analyze, from an international-relations perspective, energy issues as they evolved during the 1980s and 1990s. This book originated from that conference; its main goals are to analyze energy policy, as well as the relevant role that energy in general and energy resources in particular play in international relations. The first part of the volume includes papers related to the Gulf war and the oil markets during the 1980s. Part 2 contains papers that address issues of cooperation between oil exporters and consumers. In part 3, two papers explore the new dimensions of energy cooperation between Eastern and Western nations. Part 4 covers Mediterranean cooperation and Greece's role as a "bridge" on the European energy scene. Part 5 investigates energy policy in the European Community after the Gulf war.

184. Economic Commission for Europe. *East-West Energy Efficiency: Policies and Programmes.* New York: United Nations, 1992.

Keywords: efficiency, case studies, conservation, economics, policy

This book provides a description of the policies, international programs, grants, loan schemes, technical assistance, trade development programs, and technical standards relevant to East-West trade and cooperation in energy efficiency. Countries chosen as examples throughout the book include Sweden, The Netherlands, Germany, France, Italy, UK, Greece, United States, Japan, Lithuania, Poland, Hungary, Czech Republic, Slovak Federal Republic, Bulgaria, Russian Federation, Romania, Slovenia, and Ukraine. The third section of the book analyzes conservation policies in Eastern and Western countries and provides numerous illustrative examples. This analysis focuses on the areas of legislation, financial incentives, and technical measures. The book ends with a comparative evaluation of Eastern and Western policies, which examines market economies and economies in transition.

185. Faber, Malte. *Ecological Economics: Concepts and Methods.* Cheltenham, England:

Edward Elgar, 1996.

Keywords: case studies, economics, efficiency, environmental, fuels

Greenhouse gas emissions from human activities are widely considered to be responsible for emerging trends of global climatic change. This volume contains only one chapter related to energy issues and their link to greenhouse gas emissions. It analyzes these issues in great detail, however, and covers: anthropogenic greenhouse gases and economic development in various countries of the world; variables influencing CO_2 production; changing patterns of CO_2 emissions since 1950; examination of emissions from an economic perspective; and rates of change for CO_2 emissions for the United States, the EU, and the world. The chapter explores CO_2 emissions in Germany and the UK in relation to energy use by sector, and also analyzes issues related to future emissions trends and scenarios. The analysis focuses on: changing structures of final demand for a given growth rate; improvements in energy efficiency; changing the structure of the fuel mix between fossil fuels; the transition from fossil fuels to non-fossil fuels; and emission levels targets. In addition, the possibilities of economic change in the long run are presented, along with the problems and possibilities of achievement of CO_2 emissions reduction aims.

186. Feldman, David Lewis, ed. *The Energy Crisis: Unresolved Issues and Enduring Legacies*. London: Johns Hopkins University Press, 1996.

Keywords: energy analysis, policy, sustainability, reliability

The energy crisis of 1973-74 changed much more than the energy situation in the United States: it brought the notion of the limits of technological advancement and gave impetus to further research and developments. This volume is a collection of essays, by some of the key players of twenty years ago, that attempts to examine and draw lessons from the oil crisis experience. It explains how conceptual ideas about the crisis and government policies were formed. It also suggests how past mistakes can be avoided and how a sustainable energy path can start to be implemented today.

187. Flavin, Chris, and Nicholas Lenssen. "Policies for a Sustainable Economy." *Energy Policy* (1992): 245 - 56.

Keywords: efficiency, economics, policy, sustainability

This article looks at some of the policy challenges, such as changes in energy pricing, that must be met to achieve a renewable energy economy. The authors identify several elements needed to implement a sound energy policy, which would serve to significantly reduce carbon emissions and thereby stabilize the concentration of carbon dioxide in the atmosphere. Some of these elements are: utility reform, a major shift in current energy research and development spending, and the creation of an international renewable energy agency. The article also discusses the falling prices of renewable forms of energy generation and California's progress in this area. The authors identify five priorities for policy changes: "reducing subsidies for fossil fuels and raising their taxes to reflect security and environmental costs; reforming the electric utility industry; strengthening state

and local energy policies; increasing R & D on efficiency and renewable energy technologies; and reordering the priorities and programmes of international institutions."

188. Gaskell, George, and Bernward Joerges. *Public Policies and Private Actions: A
Multinational Study of Local Energy Conservation Schemes.* Brookfield, Vt.: Gower
Publishing, 1987.

Keywords: case studies, conservation, consumption, policy

This volume explores "the relationship between political programs and private concerns" and more specifically, "research on energy politics and energy consumers." The
first part of the book introduces a conceptual framework that serves as a base for the case
studies in part 2. The framework "is an attempt to combine policy and behavioural analysis in the evaluation of the impacts of energy conservation programmes." Consumers'
response to these programs are included as part of the analysis. The case studies, which
constitute the second part of the volume, are related to local energy conservation programs in Britain, France, the Federal Republic of Germany, the Netherlands, Sweden,
and the United States. Part 3 analyzes the issues from an international perspective where
both the "top down" and "bottom up" viewpoints are examined. This analysis highlights
"key features of both the dynamics of conservation programmes and the process of energy conservation." Part 4 explains some of the problems and limitations of applied social research; it also presents conclusions from the analysis and suggestions for policy
and future research.

189. Gay, Charles F. "Energy and the Environment: Creating New Industries." *Solar
Today* 10, 3 (1996): 16-19.

Keywords: economics, environment

This article is centered on the belief that "economic and environmental issues can be
reconciled with the help of emerging energy technologies." The United States has en-
acted some of the world's most comprehensive legislation for protecting and preserving
environmental heritage; however, some argue that these laws put intolerable burdens on
American companies. In contrast, others view these regulations as "prudent, conservative
management of the natural resources upon which our industries, and our daily lives are
built." A recent report from Shell International Petroleum Corporation predicts that, by
the year 2050, "renewable energy resources may contribute almost as much to global
energy demand as coal, oil, natural gas and nuclear combined." The cost of renewable
energy is rapidly falling: wind energy has declined from $0.50 per kilowatt-hour in 1980
to about $0.05 in 1995, thus making it competitive with conventional sources. Further-
more, it has been frequently shown that Americans support the development of non-
polluting energy technologies as a way to sustain economic growth while protecting the
environment. This article predicts that "the health of the global economy and the health
of the environment will become inextricably linked."

190. German Institute for Economic Research (DIW). *The Price of Energy.* Aldershot:
Dartmouth Publishing, 1997.

Keywords: economics, policy, sustainability

Sustainable development requires an economic system that safeguards the natural basis for life on this planet and ensures that natural resources be used more economically. One measure of achieving such a system, which is being widely discussed in Europe, is the imposition of "ecological" taxes, such as energy taxes, to make energy use sustainable. It examines the economic effects of an ecological tax reform and how such a change could lead to a significant cut in energy consumption without posing a threat to German competitiveness. The five chapters of the book have a variety of authors and cover topics such as the political battle over the concept of sustainable development; ecology and economy; measures for a climate-compatible and low-risk energy supply; ecological tax reform as a panacea or a job-killer; and Europe's arduous road to ecological tax reform.

191. Grubb, Michael, and John Walker. *Emerging Energy Technologies: Impact and Policy Implications*. Dartmouth: Dartmouth Publishing Co, Ltd., 1992.

Keywords: renewable, case studies, economics, efficiency, policy, supply and demand

This book notes that as new technologies penetrate energy markets they influence the energy business in terms of demand, supply, and policy. Four supply technologies and four demand-side technologies are reviewed. Both groups are examined through case studies that analyze the technology's "characteristics and status, its potential impact, and the factors which will determine its progress." The progress of these new energy technologies will, however, "depend heavily on the evolution of government and industrial policy towards them and broader energy issues." Part 1, entitled "Historical Patterns and Current Context," contains the following chapters: "Introduction and Historical Overview"; "The Evolving Pressures"; and "The Technological Menu," which looks at various types of technologies. Part 2, entitled "Demand-Side Technologies," includes chapters on clean and efficient cars, energy-saving strategies for domestic electrical appliances, energy-efficient lighting technologies, and building energy management. Part 3 looks at supply-side technologies such as gas turbine systems, clean coal, wind energy, and solar electricity from photovoltaics. Part 4, entitled "Synthesis," contains the following chapters: "Case Study Comparison and Impacts," "The Future Context," and "Emerging Energy Technologies."

192. Havlicek, Peter. "First Steps Toward a Cleaner Future: The Czech Republic." *Environment* 39, 3 (1997).

Keywords: case studies, environment, sustainability

There is no question that the natural environment in the Czech Republic suffered during the Communist era. Although many problems still exist, slow improvements are being made. Since being the second worst air polluter in Europe per capita in the 1980s, the government has taken steps to stabilize the situation. The issues of water pollution and soil pollution are discussed, with a close look taken at the improvements that have been made in all three areas. The article also addresses the question of the future of the environment and improvements yet to be made.

193. Holder, Jane, et al., ed. *Perspectives on the Environment: Interdisciplinary Research in Action.* Brookfield, Vt.: Avebury, 1993.

Keywords: electricity, case studies, environment, policy

This book was developed through the interdisciplinary research efforts of several experts whose work relates to environmental issues and concerns. It contains papers presented at the first IRNES conference held in 1992, in Leeds, England. In terms of energy, chapter 9, "Global Warming and Energy Policy: Burden or Opportunity?," is of particular interest: it describes the global warming response strategies of Germany, the Netherlands, and the United Kingdom. It specifically analyzes energy policies to decrease greenhouse gas emissions in the electricity sector. The aim of the chapter is to help develop better policy integration strategies that satisfy environmental and other objectives. The last part of the chapter explores the future of energy policy.

194. International Energy Agency. *Comparing Energy Technologies.* Paris: OECD/IEA, 1996.

Keywords: environment, case studies

Currently available methodologies for comparison of energy technologies are examined with suggestions for their development, for consideration by both R & D practitioners and R & D policy-makers. Two important issues are looked at: the need to review and prioritize energy technology approaches, with proposals aimed at curbing future greenhouse gas emissions, and secondly, the need in many countries to decrease resources that can be devoted to energy research, development, and technology. Actual examples of how governments go about making energy R & D choices are also presented. In an attempt to improve the process of comparing and prioritizing energy technologies, this book brings together an initial collection of work in this field of analysis, carried out under various auspices and in several countries (France, Italy, Japan, the Netherlands, the UK, and Canada), including the European Commission.

195. International Energy Agency. *New Electricity 21: Designing a Sustainable Electric System for the Twenty-First Century.* Paris: OECD/IEA, 1996.

Keywords: electricity, developing countries, sustainability, policy, energy analysis, case studies, pollution prevention, environmental impacts

This publication contains the proceedings of the IEA/UNIPEDE Conference, "New Electricity 21: Building a Sustainable Electric Future," held in Paris May 1995. The conference focused on three main areas: electricity and the environment; the impacts of growing competition in electric power markets; and the power technology needs of developing and transitional economies. The numerous papers published in this volume are divided into the following six subject areas:
1. Opportunities to increase electricity use for sustainable energy development;
2. Electric system expansion and integration to meet growing competition;
3. Power producers and global climate change issues;
4. Technology for supplying electricity in developing and transitional economies;

5. Power industry structure, regulatory policies, and technological innovation;

6. Electric technology transfer: East-West and North-South.

Although a large proportion of this volume is aimed at technical specialists, it can be of interest to readers seeking to expand their understanding of sustainability issues and current developments on the electricity field.

196. Jackson, Marilyn. *Innovative Energy and Environmental Applications*. Lilburn, Ga.: Fairmont Press, 1993.

Keywords: alternative energy, alternative fuels, case studies, cogeneration, conservation, consumption, DSM, economics, electricity, environmental, generation, policy

This extensive reference volume consists of papers presented at the World Energy Engineering Congress in 1992 in Atlanta, Georgia. The volume is divided into ten sections, each of which contains numerous chapters written by different authors. Section 1 examines environmental management in terms of: indoor air quality, monitoring for CFCs, alternative fuels vehicles, boiler technology, refuse-derived fuel, indoor air pollution, and waste conversion. Section 2, "Demand-Side Management" (DSM), deals with the power quality, bidding, process evaluations, and DSM from the perspective of a combination of utility, energy-efficient partnerships, lighting improvements, and other related topics. Section 3 discusses aspects of power generation, cogeneration, and independent power by looking at the Rhode Island Medical Center Cogeneration Plant, wood fired cogeneration, competitive bidding, retail wheeling, low grade and waste fuels, and alternative forms of energy generation. Section 4 examines energy management applications, such as: techniques for quantifying leaks in compressed air systems; systems approach in energy management; hot water case studies; innovative concepts for electrical monitoring; variables that affect energy consumption; ballast disposal; process energy reduction; thermoeconomics; window selection; and rate intervention as an alternative pricing strategy. Section 5 examines advances in lighting efficiency and applications. Section 6 briefly discusses electrical system optimization. Section 7 looks at federal energy management in terms of new directions, the role of government in the National Energy Act of 1992, energy conservation in the federal sector, and conservation strategies. Section 8, "HVAC System Optimization," presents several case studies, cost reduction initiatives, a step-by-step energy audit, and other topics. Section 9 analyzes alternative energy development, and Section Ten deals with thermal energy storage.

197. Jansen, J. C., and V. W. Buskens. "The role of sustainable energy issues in development cooperation." *Resources Conservation and Recycling* 12 (1994): 99-114.

Keywords: developing countries, economics, policy, generation

The task of providing essential energy services is an enormous challenge for the governments of developing nations, largely because the implementation of this task requires large amounts of scarce capital. The authors of this paper note that by the end of the 1980s about 25 percent of total long-term debt service payments by developing countries were due to energy loans. This proportion was expected to increase for many developing nations during the 1990s and beyond. Multilateral and bilateral development agencies are expected to provide loans to satisfy about one-fourth of the required annual energy-related investment of developing nations. The authors note that this funding level will result in macroeconomic problems that will prevent the satisfaction of prevailing energy demand trends. The main focus of this article is an analysis of the links between energy and environmentally sustainable development within the context of international agreements. It specifically deals with the implications to energy policy of the Brundtland Report, the Framework Convention on Climate Change, and Agenda 21.

198. Karekezi, Stephen. "Energy Policy Issues in Africa." *Resources, Conservation and Recycling* 12, 1-2 (1994): 4.

Keywords: case studies, DSM, environment

Despite the fact that Africa has enormous energy resources and much exploitable renewable energy potential, one large problem that exists is the insecurity of access to adequate levels of energy services. This article begins by discussing present problems, then looks at the formulation and implementation of an environmentally sound and secure energy strategy for Sub-Saharan Africa. The introduction of environmental management and DSM to the African Energy Policy and Research Network (AFREPREN) is investigated, and several measures are suggested for the promotion of renewables and DSM.

199. Kleinpeter, Maxime. *Energy Planning and Policy*. Toronto: John Wiley & Sons, 1995.

Keywords: economics, environmental, planning, policy, sustainability

The thesis of this book is that "education applied to a whole complex of interlocking problems is the master key that can open the way to sustainable development." The author notes that an important problem affecting the development of renewable energy alternatives is the lack of relevant information for engineers, technicians, decision-makers, and users. This book attempts to fill this lacuna by giving a comprehensive account of energy planning—from energy balances and forecast studies to the special features of energy planning for the developing world. Analytical models and simulation models are also covered in detail.

200. Krause, Florentin, Wilfred Bach, and Jonathan Koomey. *Energy Policy in the Greenhouse*. New York: John Wiley & Sons, 1992.

Keywords: environmental, policy, sustainability

The goal of this study is to provide guidance for the formulation of adequate energy policies to help stabilize the global climate. The first part of the volume provides background information on the energy and non-energy related factors that influence climate change. A budget is suggested for future carbon emissions that would "meet specified limits on the risks of human-induced climate change," and the emissions of other greenhouse gases are analyzed. The authors indicate how different regions and countries can contribute a fair share to solve this global problem. The last part of the study describes guidelines for the creation of an international convention on climate stabilization and sustainable development.

201. Kursunoglu, Behram N., ed. *Making the Market Right for the Efficient Use of Energy*. Commack, N.Y.: Nova Science Publishers, 1992.

Keywords: efficiency, alternative energy, conservation, economics, environmental, planning

The papers presented here provide a comprehensive analysis of the ways in which market failures may be overcome to make energy efficiency economically attractive. Regulatory and other policy initiatives by local state governments are also discussed. The generation of energy efficiency standards is considered to be a priority.

202. Levine, Mark D., Jonathan G. Koomey, James E. McMahon, Alan H. Sanstrand, and Eric Hirst. "Energy Efficiency Policy and Market Failures." *Annual Review of Energy and the Environment* 20 (1995): 535-55.

Keywords: efficiency, economics, policy, DSM, case studies

The authors maintain that the market for energy efficiency is far from perfect and provide significant empirical evidence to make this point. Their analysis points to two conditions that need to be fully addressed to enhance the effectiveness of energy efficiency policies. The first condition relates to market rejections of cost-effective energy efficiency technologies and systems. The second relates to the notion that careful evaluation of policies that promote energy efficiency clearly illustrate that benefits outweigh costs. In addition to addressing these facts, the success of energy efficiency policies is closely related to careful design, political feasibility, and quantification of policy impacts. The article examines the effectiveness of energy efficiency policies such as standards, utility demand-side management programs, and other governmental programs that have been used to overcome market failures.

203. Lowe, Ian. "Greenhouse Gas Mitigation: Policy Options." *Energy Conversion and Management* 37, 6-8 (1996): 741-46.

Keywords: environmental, economics, energy analysis, policy, supply and demand

Lowe maintains that strategies to reduce energy consumption by altering fuel or energy prices are not very effective. Experience with price signals, for example in the case of transport fuels, has demonstrated that energy demand is highly inelastic. Furthermore, energy use seems to be shaped by social factors: numerous studies show the absence of clear links between energy use and factors such as weather, climate, and changes in energy consumption over time in a particular society. In order to decrease greenhouse gas emissions it is imperative to change prevalent patterns of energy supply and use. Such change will require the alteration of social attitudes by influencing individual behavior and increasing public acceptance of regulatory measures. The author notes that significant changes in the pattern of energy use will not occur unless our understanding of the social dimensions of energy use increases significantly.

204. MacKay, R. M. and S. D. Probert. "National Policies for Achieving Energy Thrift, Environmental Protection, Improved Quality of Life, and Sustainability." *Applied Energy*

51 (1995): 293-367.

Keywords: conservation, economics, environmental, policy, sustainability

The authors note that most governments are not developing adequate responses to the depletion of fossil fuels and the widespread environmental degradation caused by the wasteful use of resources and our non-sustainable lifestyles. This article aims to identify issues that are relevant to the development of sustainable energy policies. For this purpose, different countries are classified according to two factors: their level of economic development and the availability of indigenous energy resources. The article also analyzes the influence of regional trade blocs and the level of their protectionist measures. The assumption is that developing nations tend to benefit when they join regional trade blocs. Topics include global resources; environmental issues; development as a yardstick; established and evolving trading blocs; the drive towards global sustainability; and energy and environmental policies for achieving global sustainable development.

205.Maillet, Pierre, Douglas Hague, and Chris Rowland, eds. *The Economics of Choice Between Energy Sources*. London: Macmillan, 1987.

Keywords: economics, developing countries, environmental, policy
This book contains the proceedings of a conference held by the International Economic Association in Tokyo, Japan. It is divided into the following six sections: "Methodology"; "The Energy Background"; "Policy Choices in Developed Countries"; "Policy Choices in Developing Countries"; "Environmental Problems"; and "Conclusions." It presents analyses related to the cost, role, and future prospects of energy use. Examined in detail are the role of economic theory and model-building in energy supply versus reducing energy use through the implementation of efficiency measures; the determination of adequate levels of aggregation for efficient energy planning; and financial challenges such as the difficulties of capital formation, long lead times, and the high implementation costs of several options.

206. Modl, Albert, and Ferdinand Hermann. "International Environmental Labeling." *Annual Review of Energy and the Environment* 20 (1995): 233-64.

Keywords: sustainability, pollution prevention, policy, case studies, environmental

Eco-labels can help consumers make more environmentally sound choices. Eco-labels are becoming increasingly recognized around the world, and their use could theoretically be expanded to a wide range of products and services including those of the energy sector. This article analyzes the eco-label programs of Germany, Canada, Norway, Japan, Sweden, United States, New Zealand, France, Austria, India, Holland, Singapore, Korea, Spain, the Czech Republic, and the European Union. These programs represent a wide spectrum of eco-labeling strategies that differ in focus, scope, implementation experience, origin, and goals. The article discusses the main advantages and disadvantages related to each program.

207. O'Keefe, Phil. "Trying to develop Third World energy policy—the limits of intervention." *International Journal of Environmental Studies* 50 (1996): 201-12.

Keywords: developing countries, policy, renewable, electricity, sustainability

The development of energy policy for the Third World is a complicated and challenging endeavor. This article attempts to facilitate this task by highlighting some of these difficulties and suggesting possible strategies to overcome them. The author emphasizes that although energy decisions have a significant environmental impact, these decisions are too often made by Ministries of Finance and not by Ministries of Energy or Environment. The article explores the role of energy in production, issues of energy and equity, the emerging nature of energy markets (with a focus on electricity and the role of renewable energy), and the potential role of energy institutions as agents for sustainable energy development.

208. Peirce, William S. *Economics of the Energy Industries*. Westport, Conn.: Praeger, 1996.

Keywords: economics, electricity, nuclear, generation, policy, alternative fuels

Peirce provides a broad perspective on the economic issues of several energy industries. His main purpose is to introduce the reader to the field of energy economics in order to facilitate and enhance understanding of energy policy issues. The book is divided into four parts. First, it defines economic concepts and provides an overview of changes experienced by the energy markets since the oil crisis of the 1970s. The second part explores the economics of the coal, oil, and natural gas industries. This is followed by an analysis of the use of electricity, nuclear energy, and alternative sources of energy generation. The volume concludes by summarizing several public policy questions related to the energy industries analyzed.

209. Rogoff, Marc J. *How to Implement Waste-to-Energy Projects*. Park Ridge, N.J.: Noyes Publications, 1987.

Keywords: economics

This book presents information on what steps must be undertaken to implement a waste-to-energy project. It is a very "practice-oriented" book that gets into all aspects of laws, regulation, siting, economics, and marketing. Chapter 1 deals with the solid waste disposal problem and the trends towards waste-to-energy. Chapter 2 centers on project implementation concepts, and more specifically in terms of developing project teams, risk assessment, the implementation process, and public information programs. Chapter 3 explores waste-to-energy technology in terms of mass burning, modular combustion, refuse derived fuel (RDF) systems, and fluidized bed systems and, then compares these technologies. Solid waste composition and quantities are discussed in chapter 4, along with characterization and composition studies of waste, heating value, and quantities. Chapter 5 looks at waste flow control and mechanisms. Chapter 6 examines the selection of facility cities and the site selection and screening process. In chapter 7, energy and materials markets are discussed. Chapter 8 covers the permitting of waste-to-energy facilities and the Dioxin issue. Chapter 9 focuses on procurement of waste-to-energy systems, procurement approaches and procedures, preparation of requests for proposals, and proposal evaluation. The final chapter looks at the ownership and financing of a waste-to-energy facility in terms of ownership alternatives, financing options, key participants in resource recovery financings and steps in bringing the bond issue to market.

210. Russel, Jeremy. *Energy and Environmental Conflicts in East/Central Europe: The Case of Power Generation*. London: Royal Institute of International Affairs and World Conservation Union, 1991.

Keywords: case studies, developing countries, policy, sustainability, reliability, generation

The collapse of the Soviet Union has resulted in an energy policy challenge for Central and Eastern European countries. These nations need to develop new energy strategies to ensure their supply of electricity and to substitute for energy imports that used to be provided by the former Soviet Union. This challenge is further complicated by the need to implement energy policies that are environmentally sound. This report explores the advantages and limitations of the energy options that are available for Central and East European countries.

211. Schmidheiny, Stephan. *Changing Course: A Global Business Perspective on Development and the Environment*. Cambridge, Mass.: MIT Press, 1992.

Keywords: economics, sustainability, case studies, policy

Business and environmental sustainability have often been perceived as opposing forces. *Changing Course* aims to prove that both forces can, and should, be linked. The author illustrates this point by providing numerous case studies. In terms of energy issues this volume provides one chapter on energy from a business perspective and a concise appendix on priorities for a rational energy strategy.

212. Shaw, Lennard, and P. M. S. Jones. *Policy and Development of Energy Resources*. New York: John Wiley & Sons, 1984.

Keywords: developing countries, economics, environmental, fuels, policy

There is much debate as to whether renewable forms of energy will be implemented on a large scale in the near future. This book, published some ten years following the "cheap oil" era, suggests that this is indeed the case and that there will be a natural transition stage between the present forms of energy in use and renewable ones of the future. In the meantime, sociopolitical policy, it is argued, must thoroughly take advantage of all promising and informative progress and provide support in the form of funds. Designed to inform engineers, planners, politicians, and scientists, the book takes a broad look at the energy situation facing us today, not only with reference to individual source development, but also with regard to issues like economics, politics, and environmental considerations. It insists that "the world is not short of energy, only the unfailing foresight to make it available in the right form and cost, at the right time and place." Sections in the book include: "A Fuel's Policy into the Twenty-First Century," "Outside the Oil Economy: Rural Energy for Developing Countries," "Trade Unions and Energy Policy," "Financing World Energy Projects," "Environmental Considerations," "Coal Supplies," "Oil and Gas," "The Nuclear Contribution," "Renewable Energy Sources in Perspective," "Energy Storage," and "Energy Conservation in Transport, Manufacturing, and Buildings." It should be noted that, although this book is quite comprehensive, it is somewhat dated in a few of its arguments.

213. Smeloff, Ed, and Peter Asmus. *Reinventing Electric Utilities: Competition, Citizen Action, and Clean Power.* Washington, D.C.: Island Press, 1997.

Keywords: economics, policy, case studies, electricity, generation, nuclear, environmental impacts, sustainability

Alternatives to the prevailing organization of the electric services industry in the United States are currently being debated and implemented. Smeloff and Asmus analyze, in great detail, the experiences of the Sacramento Municipal Utility District (SMUD) and several other U.S. utilities. They use these case studies to illustrate the problems that utility restructuring efforts have encountered and possible ways to enhance the sustainability of future efforts. Utility restructuring efforts are complicated by the onerous legacy of unproductive and environmentally damaging forms of energy generation and by the challenge of developing effective renewable energy and energy efficiency initiatives. The authors specify that in order to ensure a more egalitarian and environmentally benign electricity supply several important factors must be incorporated into the decision-making process: creative leadership, open and competitive markets, and active citizen participation. These factors are essential for the development of a more sustainable, competitive, and fairer system of electricity supply.

214. Smith, Zachary A. *The Environmental Policy Paradox.* Englewood Cliffs, N.J.: Prentice-Hall, 1995.

Keywords: conservation, efficiency, nuclear, policy, renewable, transportation

This book deals with many aspects of environmental policy, and those related to energy are addressed in Chapter 7. The chapter begins with an analysis of current environmental policy, then presents the history of energy use, from its growth during the industrial revolution to the importance of oil during World War I. Personal consumption in the United States; OPEC and the oil crisis; the development of nuclear power; and the development of a national energy policy are discussed. The author looks at various sources of energy such as coal, oil, natural gas, geothermal energy, and nuclear power and analyzes their environmental impacts. He then considers renewable forms of energy in the same terms: hydropower, solar power, wind power, and biomass. Finally, this chapter deals with conservation and energy efficiency and makes suggestions for the future. The ecological benefits of energy conservation in homes and buildings, in transportation, and in industry are well analyzed, as are the current obstacles to conservation that need to be overcome through future policy initiatives.

215. Swezey, Blair G., and Yih-Huei Wan. "The True Cost of Renewables." *Solar Today* 9, 6 (1995): 30-32.

Keywords: renewable, economics, environmental

There is great controversy over the actual cost of renewable energy resources. Some believe that an increased use of renewable energy in a country's power sector would translate into unaffordable costs for the nation's electricity rate-payers. Others, however, believe that a modest yet continuous growth path of renewable energy resource development would not cost the nation more than projected electricity market costs for coal-based power generation, and there would be many environmental benefits as well. This article looks at the true cost of renewables, energy subsidies (such as massive public subsidies for coal and other fossil fuels), environmental impacts, the reliability of renewable energy systems, and the impact of electricity competition. At the end of the article, a separate section, "Another Perspective" by Ronal W. Larson and Ronald E. West, offers an "optimistic" perspective on the costs of renewables, based in part on information gathered in preparing the 1992 ASES economics white paper.

216. Teixeira, Maria Gracinda C. *Energy Policy in Latin America*. Aldershot, England: Ashgate Publishing Co., 1996.

Keywords: alternative energy, case studies, economics, environment, hydroelectric

This book looks at the role of the Brazilian Electric Power Sector in the "development" of Amazonia for its main industrial projects, the large hydroelectric dams, and their impact on the Amazon forest and local communities. It also addresses the links between the planning of hydroelectric schemes in Amazonia and the development of mining industries in terms of their effects on local and regional economics, environments, and populations. The controversial Brazilian dam of Tucuri is used as an example and compared to the projected scheme of Cachoeira Porteira, which leads to discussions of the social and institutional actors involved in the construction of large dams in Amazonia and the evolution of the role of national and international financial bodies in promoting both industrial enterprises. Emphasis is placed on the fact that the political economy of resource utilization in Amazonia has been a sequential state-driven activity to reinforce central government power over the regions, and has led to the exploitation of natural systems and local populations. It is for this reason that the two eras of electrical power are compared (the Tucuri Dam and the projected dams of Cachoeira Porteira). Because this is not a technical book, no technical solutions are proposed for the political problems posed by the carrying out of these schemes. Rather, an actor-oriented approach is used, combined with a structural and political economy analysis. Also emphasized are social and political responses and how local, national, and international structures have interacted to promote this public policy until the present.

217. Tester, Jefferson W., David O. Wood, and Nancy A. Ferrari, eds. *Energy and the Environment in the 21st Century*. Cambridge, Mass.: MIT Press, 1991.

Keywords: environmental, economics, end-use, policy, sustainability

This book contains the proceedings of a conference held at the Massachusetts Institute of Technology (MIT). It begins with papers written by government officials, industrialists, and academics who looked into environmental science, demographics, economics, policy, and energy technology issues. Following these, some end-use areas were discussed, including transportation systems, industrial processes, building systems, and electric power systems. Each one of these areas was discussed in relation to the environmental impacts of current technologies and means of reducing these impacts. Suggested solutions were conservation, improved energy efficiency, re-engineering of processes and materials, fuel switching, and the use of new energy supply technologies. Both developed and developing countries were considered in this discussion about technology and economic concerns, critical policy options, and appropriate research and development priorities. Other issues that were addressed were economics and policy and advanced energy supply technologies such as solar photovoltaic, hot dry rock geothermal, magneto-hydrodynamic, and fusion energy. Different perspectives were presented on "policy strategies for the efficient and equitable management of energy resources to create a sustainable global environment."

218. Van Wees, Mark, and Ad Van Wijk. "The Assessment of the Long-Term Prospects of Energy Technologies in the Netherlands." *Energy Conservation Management* 37, 6-8 (1996): 712-722.

Keywords: energy analysis, environmental

This article covers the actions of a program called SYRENE, which was commissioned by the Netherlands Agency for Energy and the Environment (Novem) to address the issue of setting R & D priorities. It is claimed that in order to set energy R & D priorities, it is necessary to assess all current information regarding new or improved technologies and their value for the future. The fairly in-depth analysis that was done regarding the characteristics of energy technologies on different levels is well presented. Conclusions to the value of this program are put forward at the end of the article, especially in terms of the "relative long-term prospects of energy technologies with a focus on the reduction of CO_2 emissions."

219. Vine, Edward, Drury Crawley, and Paul Centolella, eds. *Energy Efficiency and the Environment: Forging the Link*. Washington, D.C.: American Council for an Energy-Efficient Economy, 1991.

Keywords: efficiency, conservation, energy efficient buildings, environmental, planning

Leading researchers, program analysts, and policy-makers put together papers for the ACEEE (American Council for an Energy Efficient Economy) 1990 Summer Study on Energy Efficiency in Buildings, which resulted in this book. It examines the links between energy conservation and environmental issues like global warming, air pollution, acid rain, and ozone depletion and tries to demonstrate how energy efficiency in the home, in the office, or with regards to transportation can alleviate such problems. It also looks at how planners can take into account environmental externalities as they select and use resources. A wide range of topics are covered in this book, such as: the public conceptions and international perspectives of global warming; efficiency and regulatory policy options; environmental externality costs; integrated energy and environmental planning; trees and landscaping; and urban heat islands. Several articles also look closely at carbon dioxide in terms of its effect on global warming.

220. Wells, Donald T. *Environmental Policy: A Global Perspective for the Twenty-First*

Century. Upper Saddle River, N.J.: Prentice-Hall, 1996.

Keywords: alternative energy, conservation, economics, environmental, policy, renewable

Energy policy reveals some basic problems in understanding environmental issues. Chapter 8 of this book focuses on the effects environmental objectives should have on energy production and use. This comprehensive approach allows us to see both environmental problems and energy needs in the larger ecological context within which they occur. The following topics are covered: energy consumption, parabolas, and a high cultural discount rate; energy rotations; per capita GNP, energy requirements and greenhouse gas emissions in selected countries; and alternative energy sources in an environmental perspective, such as biomass energy, wind energy, geothermal energy, and other potential energy sources. Energy conservation policy with an environmental perspective in terms of market forces, subsidies and regulations, public funding of research and development, and a "fledgling environmental policy on energy" are also explored.

221. Wilbanks, T. J. "Implementing Environmentally Sound Power Sector Strategies in Developing Countries." *Annual Review of Energy* 15 (1990): 255-76.

Keywords: environmental, developing countries, economics, electricity, policy

We have long been challenged to deliver electricity services to developing countries in forms that are consistent with their levels of economic development and parallel to their social fabrics. Additional effort is required to provide such services in ways that do not cause damage to the environment or weaken the social fabric of North-South relationships. This task is made all the more complex as a result of several issues: quickly growing populations in these countries; the stress that the global environment is already under; and the growing cost of energy from petroleum and natural gas for electricity generation while coal, a major cause of damage to the environment, is still quite abundant and widely used. These related issues serve as the basis for discussion in this paper, as it is one of the most important predicaments of our time. Some main issues and options related to the implementation of environmentally responsible strategies for the development of electric power systems are addressed, and it is made clear that the challenge is more complex than simply knowing what to do, "whether the choices involve resource and technology development or impact mitigation strategies."

222. Williams, Robert H., and Gautams Dutt. "Future Energy Savings in U.S. Housing." *Annual Review of Energy* 8 (1983): 269-332.

Keywords: case studies, conservation, consumption, policy

Energy prices in the United States increased drastically in the 1970s: oil costs tripled, natural gas doubled, electricity went up by a third, and energy costs overall doubled. Due to these increases, residential energy consumers, who make up one-fifth of all energy use in that country, had to cut back significantly on their use. These cutbacks amounted to 10 percent between 1978 and 1980. In the first section of the article, an estimate of the potential for technically feasible and cost-effective energy conservation in single-family housing by the turn of the century is provided. Ground rules are then outlined which were used in measuring technical feasibility, and the major technological options that provide the basis for the estimate of the energy conservation potential are surveyed. The article concludes by discussing policy issues, advancing suggestions promoting techno-logical opportunities, and formulating strategies.

APPENDIX

California's Energy Crisis: A Sign of Things to Come

When completing this bibliography, we assumed that identifying the potential for preventive approaches for the production, distribution and use of energy might spur attempts to make modern ways of life more sustainable. In the case of electricity and natural gas, where a costly infrastructure for distribution is required, publicly regulated monopolies were created and developed to ensure an adequate and cost-effective supply of energy to customers in their jurisdictions. The efficiency with which customers converted this energy in order to obtain the services they required was of no concern to these utilities, nor was it a concern for the manufacturers of the devices providing these services. As a result, a great deal of technological effort ensured that the production and distribution of electricity and gas was technically highly efficient and economically reasonably cost-effective. In contrast, energy end-use efficiency was low in comparison to what was technically and economically possible. However, there were important variations. Some countries with high energy costs were able to produce a unit of GDP with half the energy input than that required in other countries where energy was relatively inexpensive.

This discrepancy led to the realization of a possible alternative energy strategy. Its focus would not be primarily on supply, in the recognition that there were two ways of meeting a growing energy demand. The conventional approach only considered alternative options for producing ever-larger quantities of energy. Warnings that the exponential growth of energy demand could not be indefinitely satisfied were ignored. The alternative strategy recognized that a growing demand for energy could be met differently. Generating a kilowatt-hour of electricity, for example, would have the same effect as negatively generating it by improving the efficiency of energy end-use. The negawatt revolution had begun, thus transforming the energy business in jurisdictions such as California.

There was an unresolved problem. Frequently compelled by governments, utilities had heavily invested in nuclear power, which turned out to be a monumental failure of energy planning. Instead of delivering power "too cheap to meter," it failed to deliver power cost-effectively, not even considering the many other problems including the disposal of radioactive waste. Many utilities had run up huge debts, and some of the largest customers sought a way out of having to help pay for them. They therefore lobbied for wholesale competition, arguing that energy markets had to resolve the problems.

In the meantime, the potential of the negawatt revolution was being tapped by Integrated Resource Planning, which considers both supply and energy efficiency options for meeting energy demands. In many jurisdictions, Integrated Resource Planning became an integral part of the regulatory framework within which utilities operated. It encouraged and rewarded those utilities which rethought their business. Instead of focusing on demand, becoming an energy service provider was not only more lucrative but also beneficial to all parties. The idea was a simple one. Customers were approached to share the costs and benefits of the improvement of energy end-use efficiency through retrofitting the systems that delivered the desired services. Integrated Resource Planning incorporated wholesale competition by the utilities inviting industry to bid on supply generated by co-generation, for example. However, it would not protect large customers from taking on their fair share of the enormous debts run up as a consequence of conventional energy strategies.

Events in California took everyone by surprise, including the authors of this bibliography. Would the window of opportunity provided by natural gas as a transition fuel to a non-fossil-fuel-based energy system be much smaller than anticipated? Several aspects must be considered. First, it has been suggested that gas wells in North America have peaked. As the cost of natural gas is rising sharply, the conversion back to oil has begun. However, oil wells in North America have peaked as well. The usual economic argument that, as the price of gas and oil shoots up, an additional supply will come within technological and economic reach is undoubtedly true, but the increase in supply is likely to be very small in relation to annual consumption levels.

A second aspect to be considered is that converting the present energy infrastructure use from renewable resources itself will require a great deal of energy. For example, a solar panel requires a considerable energy investment to produce and operate it. If this gross energy requirement is substantial in relation to the energy delivered by the panel over its lifetime, the increase in energy supply from solar panels may be very modest. This is not an economic question but a thermodynamic one. It is essential to carefully examine how renewable the renewable energy sources really are. This is determined by the total investment of energy required to deliver the energy produced by the system from renewable resources. To convert the current energy system will require a huge investment of energy, and the energy payback significantly depends on the actions we take in the immediate future. As a footnote, we must end the confusion between energy sources and energy carriers. Electric cars are zero emission vehicles only if the electricity is produced from renewable sources, including the energy required to deliver it. The same is true for energy obtained either directly from hydrogen or from hydrogen-based fuel cells. In sum, living off the interest from the renewable energy capital available to us requires that it supplies our direct needs and that it also covers the energy required to produce this energy. This is true for solar, wind, and tidal sources. For those who place their hope in nuclear power, it should be remembered that some studies claim that a nuclear plant requires a greater energy investment than it returns in its lifetime.

In sum, it is time we begin to take into account the thermodynamic limits within which we live. A society can neither create nor destroy energy but only transform it. If we are to use the window of opportunity to transform our energy infrastructure, it is essential that we shift the focus away from gross energy production to net energy availability. If a new energy option requires more energy to produce and operate it, proceeding with it will in effect reduce the net energy available to society. On the end-use side, we need to continue what energy analysis began: the primary focus should be on the services required to sustain a modern way of life, and these should next be examined from an energy perspective to ensure that they can be delivered as efficiently as possible with the least interference with human life and the biosphere. Events in California (increasingly replicated elsewhere, including Alberta and Ontario) demonstrate that we are far from coming to terms with the issues and even further from a clear strategy to make use of the narrow window of opportunity in which to create a sustainable energy future.

The third and final aspect is what we will call the Easter Island effect. Our window of opportunity to a sustainable energy future may be closing more rapidly than originally expected. Yet the voices warning us go largely unheeded. We cannot help wondering whether the situation on Easter Island just prior to the collapse of that advanced civilization was not similar to our own. Were people sounding the alarm dismissed as doomsday prophets? Did people expect that somehow, somewhere an alternative to the increasingly scarce wood supply would magically appear at the last moment? Here we enter into a debate on the cultural plane, namely, the ability of a community to clearly recognize what it is doing to itself and to mobilize its resources if and when a crisis is discerned. An economic fundamentalism that places almost limitless confidence in energy markets could turn out to be a primary ingredient in creating an Easter Island effect. Such a fundamentalism is rooted in the myths (in the sense of cultural anthropology) that obscure our ability to see what is happening. This appendix is our attempt to read the signs of our time as they began in California, which adds the dimension of time to the main body of this bibliography.

223. Bartlett, Albert A. "An Analysis of U.S. and World Oil Production Patterns Using Hubbert-Style Curves." *Mathematical Geology*, 32, 1 (2000).

Keywords: oil reserves, forecasting

Three-fourths of all the oil reserves in the United States have already been mined through 1995, and thus oil recovery is in serious decline. World oil production is expected to peak in 2004 and then begin to decline as well. The author presents a few other scenarios based on differences in the demand for oil or finding surprising reserves. The conclusion however is the same: the end of the oil era is coming soon and can only be delayed by a few years.

224. Campbell, C. J. *The Golden Age of Oil, 1950-2050: the Depletion of a Resource.* Dordrecht: Kluwer Academic Publishers, 1991.

Keywords: oil reserves, oil industry, forecasting

This is a fascinating look at the oil industry and the world that it has wrought. Its essential message is the same as the one articulated in his later, much more analytic work *The Coming Oil Crisis* (*see* no. 225). Campbell introduces us to all aspects of oil production and trade in an insightful manner that reflects his vast knowledge of this topic.

225. Campbell, C. J. *The Coming Oil Crisis.* Brentwood, England: Multi-Science Publishing Company and Petroconsultants, 1997.

Keywords: oil reserves, forecasting

The twentieth century has been largely shaped by oil and our economies continue to be critically dependent on this resource. Large aspects of Western foreign policy are dictated by the need to keep access to this resource secure. In large measure, critical aspects of oil's role cannot be substituted for by other energy resources, at least not for the foreseeable future. We will need oil to power our jet airplanes, large earthmovers, and so on, not to mention the thousands of synthetics that are derived from it for the manufacture of everything from textiles, plastics, and pharmaceuticals to fertilizers. The question of how much oil remains to be found and for how long this resource can continue to meet the growing demands for it (especially with rapid development taking place in two of the world's most populous nations, China and India) is thus of crucial importance. This is precisely the task that Campbell has set for himself in this book.

This is a thoroughly interdisciplinary and well-informed book. Campbell's tremendous experience in the world of oil is a great asset, as is his access to the extensive database of Petroconsultants. Campbell has been involved in every aspect of oil production, from searching for it and finding it, to management and consulting in the oil business. The message he brings us is not a happy one, but nevertheless one that we should have expected, only not so soon. According to Campbell, the mid-point that corresponds to peak production of oil is to come within the first ten years of this millennium. Thereafter, production rates will begin to decline steeply until we are effectively out of oil by 2050 or so. All this is based on his estimation that the ultimate oil reserves amount to 1,800 billion barrels of which 1,600 billion barrels have already been found, with only 200 billion remaining to be found.

Campbell offers several scenarios of how the world is going to respond to a dwindling oil supply. The picture is not a pretty one. Massive collapse in the economies of the developing world are a given, but the developed world is not going to escape the consequences either, nor are the oil producing countries. War over oil and the possibility of the West occupying oil-producing countries is also not ruled out. In other words, the world will become a very different place, with a smaller population. In Campbell's words, "the transition [to a post oil world] will be difficult, and for some catastrophic, but at the end of the day the world may be a better and more sustainable place."

226. Campbell, C. J., and J. H. Laherrère. "The End of Cheap Oil." *Scientific American* (March 1998).

Keywords: oil reserves, forecasting

An oil crunch is just around the corner, the authors warn us, and this crunch will not be temporary like the one that came in 1973. The 1973 oil shock was motivated by politics at a time when there was actually plenty of oil around. This time around, however, the crunch is coming because we are running out of this crucial resource. The two authors have extensive experience in the oil industry around the world and have actively engaged in oil exploration for the giant oil companies. They are also aware of the many different political machinations that lie behind the "don't worry there's plenty of oil left" forecasts that are regularly made by oil companies and government agencies. The authors are in a good position to criticize such forecasts and point out their various shortcomings, which they do clearly and plausibly. The criticism of industry forecasts focuses on three critical errors that they make. The first of these is that they are based on a distorted estimate of reserves; the second mistake is the assumption that production will remain constant; and lastly the assumption that the last barrel of oil can be pumped from the ground just as quickly as the barrels coming from the wells today, is also erroneous. The authors correctly point out that the rate at which any well can produce oil always rises to a maximum, and then, when about half the oil is gone, begins falling gradually back to zero. Thus, from an economic perspective, when the world runs out of oil is not directly relevant; what matters is when production begins to decline. Beyond this point, oil prices will rise unless there is a commensurate decrease in demand. Using several different techniques, the authors conclude that this decline will begin before 2010.

227. Crow, Patrick. "U.S. Oil and Gas Industry Has High Expectations from Bush Administration on Energy Issues." *Oil and Gas Journal* 99.7, (February 12, 2001).

Keywords: oil reserves, economics, policy

The greatest challenge facing the new U.S. administration under President Bush and Vice President Cheney is an impending energy crisis. There is considerable hope in the oil and gas industry that the condition can be greatly ameliorated by the expected policy proposals of the new administration. The industry expects that the new proposals will be much more amenable to streamlining more exploration and production activity. Currently, the President must take quick action on California, which has plunged into a dark electricity crisis. Policies are also expected to reduce natural gas prices which are at an all time high and because of which many U.S. homeowners will be paying as much as 70 percent more in heating bills by the time the 2000/2001 winter ends. Heating oil and gasoline prices are also very high. Farmers are planning on cutting their corn crops this spring because high natural gas prices have made energy for irrigation too expensive and the use of nitrogen based fertilizer too uneconomic. One of the key expected developments is going to be the opening of the Alaska National Wildlife Refuge (ANWR) coastal plain to exploration and development. The president has already hinted that the ANWR is going to play a crucial role in any energy policy plans put forth by his administration. This action is not expected to pass without any opposition because environmental groups are poised for action to protect this environmentally sensitive area; moreover, the general public's attitude toward developing this area has been generally negative, at least in the main states. In Alaska there is some support for the proposal as it is expected to provide an economic boom. There is opposition from Canadians as well since most of the proposed 1,800 mile pipeline that will be built to transport any oil discovered here will pass through Canadian territory. Again, most of this territory is environmentally sensitive and most of it is in the migration paths of various animals. In associated articles in the same issue, the California situation, and the pros and cons of opening ANWR to development are discussed in greater detail.

228. Duncan, Richard C. "The Peak of World Oil Production and the Road to the Olduvai Gorge." *Pardee Keynote Symposia.* Reno, Nev.: Geological Society of America, 2000.

Keywords: oil reserves, forecasting, economics

This paper develops the provocative but plausible theory that Industrial Civilization has a life expectancy that is less than or equal to one hundred years, from 1930 to 2030. It is based on a correlation of world population data with energy data. Given the peak in world oil production, the author predicts a decline in energy production per capita that will bring it down to its 1930 value by 2030. At this point any number of factors could be cited as the causes of the collapse. Whatever the causes, the author believes that they will be strongly correlated with an epidemic of permanent blackouts of high-voltage power networks around the world. In the author's words, "when the electricity goes out, you are back in the Dark Age. And the Stone Age is around the corner."

229. Faber, Malte, Reiner Mansteten, and John Propps. *Ecological Economics: Concepts*

and Methods. Cheltenham, England: Edward Elgar, 1996.

Keywords: pollution, economics, policy, fuels

Chapter 14 of this book is one of the most comprehensive accounts of the environmental consequences of using fossil fuels. The authors give a fine analysis of the dilemma posed by the fact that almost all economic progress made in this century can be attributed to the widespread use of fossil fuels, but now it appears that the greatest danger faced by humankind, that of global climate change, is also attributable to fossil fuel use.

230. Fleay, Brian J. *The Decline of the Age of Oil.* Annandale, Australia: Pluto Press, 1995.

Keywords: oil reserves, forecasting, Australia, policy, economics

World oil production is fast approaching its peak and an inevitable decline is around the corner. What this means for the world economy in general and for Australia in particular is explored in detail in this well-informed and well-researched study. The analysis is guided by the energy theory of value developed by H. T. Odum and Robert Constanza. This new approach to economics sees all value as ultimately derived from nature, unlike neoclassical theory where value is founded primarily in human relationships through exchanges in the marketplace. This new theory does not displace contemporary economics, rather it reintegrates it within its proper environmental context. From an energy perspective, all economic activity is ultimately driven by energy flows from nature. This emerging model says the human economy is a thermodynamically open system embedded in the environment and depends on a net inflow of energy, natural resources, and other services from the environment. Value ultimately has its source in natural resources in direct proportion to the degree of concentration (or order) or complex structure stored in those resources and in inverse proportion to the physical energy cost of finding and extracting that order.

From this perspective, oil has a very high value, and its demise is going to have critical consequences. In essence, denied of the prospect of profuse oil use, the countries of the developing world will suffer the most, with none of them having the possibility of ever achieving the living standards of the West. Australians, along with the West will face drastic reductions in per capita wealth. The book suggests many policy options that Australia should deploy urgently. The suggestions are just as relevant for the rest of the developed world.

231. Gever, John, Robert Kaufman, David Skole, and Charles Vörösmarty. *Beyond Oil: The Threat to Food and Fuel in the Coming Decades.* Cambridge, Mass.: Ballinger Publishing, 1986.

Keywords: oil reserves, forecasting, agriculture, economics, policy

This book is a project of Carrying Capacity Inc., which was formed to analyze the carrying capacity of U.S. resources and how to move toward a permanently sustainable way of life in the United States, and over time, the world. In 1982 Carrying Capacity Inc. asked the Complex Systems Research Center (CRSC) at the University of New Hampshire to model the energy future of the United States and, having done that, to project its effects on the nation's agricultural system. In their own words: "The CSRC investigates problems stemming from interactions between humans and their environment. Rather than breaking down natural and global problems into isolated pieces and studying each piece separately, the Center uses mathematical modeling and policy analyses to study them as systems. Societies and national economies, inappropriate subjects for controlled experiments and physical testing, are best studied through computer-based mathematical modeling." The current study comes to the following conclusions. Energy supplies, contrary to popular belief, are going to dwindle in the very near future. By 2005 at the very latest, it will take more energy to explore for U.S. oil and gas than the wells will produce. Even though old wells will continue to be exploited beyond this date, U.S. oil will be virtually exhausted by 2020. Since oil and gas comprise 75 percent of U.S. fuel use, it is extremely unlikely that the supply of alternative fuels or the nation's energy efficiency can be increased quickly enough to completely offset the effect of declining oil and gas. A long-term downturn in U.S. gross national product is thus a likely prospect starting in the 1990s. The study predicts that since U.S. agriculture is so heavily oil-dependant it is vulnerable, to the extent that early in the new millennium the United States could lose its ability to be a net exporter of food unless there are dramatic increases in the energy efficiency of agricultural production. All of these effects are going to be felt with increased force by the rest of the world which, since it is not as affluent as the United States, will be less able to bear the brunt of the fuel crisis.

The study concludes with several policy recommendations. Gaining control of and stabilizing population is of primary importance, with an eventual eye toward reducing population. Fuel taxes and import duties must be imposed rapidly to reduce the demand for fuels and increase energy efficiencies. Development of cogeneration capacity is also urged so that the heat produced by electricity generation is not wasted. Energy conservation through efficiency measures is a major priority as well, and the federal government is urged to set an example by taking efficiency initiatives. Lastly, investment in renewable fuels and the development of less energy intensive and more sustainable agricultural practices is also urged.

232. Gever, John, Robert Kaufman, David Skole, and Charles Vörösmarty. *Beyond Oil: The Threat to Food and Fuel in the Coming Decades. A Summary Report.* Washington, D.C.: Carrying Capacity Inc., 1986.

Keywords: oil reserves, forecasting, agriculture, economics, policy

As the title suggests, this is a summary report of the study described in *Beyond Oil.* An additional feature found here that is not part of the original study is a running commentary by Carrying Capacity Inc. on the various parts of the report along with suggestions for further reading. The commentary tends to be a bit more optimistic than the report itself, especially with regard to what can be achieved through efficiency measures and through the rapid deployment of renewable energy technologies. The commentary provides a lot of useful data on how much energy can be saved by switching to energy efficient devices that are already on the market—from lighting systems to refrigerators. The data on the energy and economic payback ratios of solar photovoltaics is updated as well.

233. Hall, Charles A. S., Cutler J. Cleveland, and Robert Kaufmann. *Energy and Resource Quality*. Boulder: University of Colorado Press, 1992.

Keywords: fuels, economics, energy analysis

This is an extremely good introduction to the inherent qualities of different energy resources and the efficiency with which they can be converted into currencies. Of particular importance is a description of the energy effort required to recover resources. This explains the "peak and decline" effect associated with certain resources such as oil. The energy effort required to pump up the first half of the oil in a field requires relatively low effort, but once the halfway point is reached, the recovery is said to have peaked as thenceforth the energy effort begins to increase for every barrel of oil recovered. At some point in time the effort required to recover a barrel of oil begins to approach the amount of energy contained in that barrel of oil. At this point in time the well effectively becomes an energy sink and it does not make any economic or energetic sense to continue pumping it. From this perspective, oil fields in all of the United States have already peaked and are now in decline.

234. Hanson, Jay. "Energetic Limits to Growth." *Energy Magazine* (Spring) 1999.

Keywords: oil reserves, economics, forecasting

The coming peak in global oil production is going to bring an end to the consumer society because no other energy resource can replace conventional oil that is currently our biggest provider of net energy. Net energy, in turn, is the pre-condition for all other resources. The author provides cogent reasons to explain why natural gas, coal, nuclear power, fuel cells, etc., cannot match oil for versatility, cost, ease of extraction, and transportability. A good critique of conventional economics for ignoring the energy issue accompanies this article. Hanson lambastes economists for seeing the availability of oil only in monetary terms, i.e., oil scarcity will drive the price of oil up making it economically viable to search for oil in deeper and currently economically nonviable regions. What the economists ignore, says Hanson, is that it takes energy to find energy. At a certain point in time, digging deeper for oil, or trying to produce it from unconventional sources does not make any sense because the amount of energy required to do so equals the amount of energy in the found oil, thus making it a net energy sink. Thus there are very real energetic limits to growth.

235. Ivanhoe, L. F. "Get Ready for Another Oil Shock!" *The Futurist* (January/February 1997).

Keywords: oil reserves, economics, forecasting

Petroleum consultant Ivanhoe warns that we have found almost all the oil there is and we are continuing to use it up rapidly. There will be less oil entering the world markets in the very near future—somewhere between now and 2010, but no later. The global price of oil after the supply crunch should follow the simplest economic law of supply and demand: There will be a major increase in the price of crude oil and that of all other fuels, accompanied by hyperinflation, rationing, and other stringency measures. The author suggests that after the economic implosion, many of the world's developed societies may look like today's Russia. The United States may be competing with China for every tanker of oil, with the Persian Gulf oil exporters preferring Chinese rockets to American paper dollars for their oil. Ivanhoe says that if stringent measures (such as a substantial increase in the tax on oil and other steps to put the planet on an energy diet) are not undertaken immediately, a major negative change in our lifestyles will occur.

236. Ivanhoe, L. F. "Future World Oil Supplies: There Is a Finite Limit." *World Oil* (October 1995).

Keywords: oil reserves, forecasting

The question is not whether, but when, world crude oil productivity will start to decline, ushering in the permanent oil shock era. Ivanhoe sees this decline as occurring in the first decade of the new millennium. For what may happen after this decline and what we should do about it, see the previous article from the same author.

237. Jackson, T., and M. Oliver, eds. "The Viability of Solar Photovoltaics." *Energy Policy* 28, 14 (November, 2000).

Keywords: solar, photovoltaics, renewable

This important issue of *Energy Policy* is meant to not only update us on the latest developments in solar photovoltaics (PVs)—a task that it performs admirably—but also to urge for a faster transition to this technology. Economies driven in the main by hydrocarbons will suffer in the near future because of rapidly depleting resources and because of the central role these resources play in global climate change—a fact well acknowledged by scientists and, with greater public awareness, soon to lead to governmental action.

Both the economic and energetic payback times for PVs are consistently improving, though they remain behind other more conventional energy generation technologies. This situation can be improved through governmental intervention to encourage more research and development in this area as well as to provide incentives for their adoption. The more expensive PVs would then rapidly fall in price and their energetic payback time would improve as well if they were being produced on a mass scale rather than the piecemeal way they are put together now by scientists and engineers. The editors feel that market forces alone should not be the decisive factor in determining PV use. They are convinced that an energy crunch is imminent and those countries that have not diverted a significant proportion of their energy sector to PVs will suffer economic disasters.

238. Khawlie, M. R. *Beyond the Oil Era? Arab Mineral Resources and Future Development.* London: Mansell, 1990.

Keywords: oil reserves, economics

The author, a professor of geology at the American University of Beirut, warns of the calamitous effect that declining production of oil would have on the economies of the Arab states. This is because these economies are so centrally configured around oil. He warns that reserves were already declining in 1990 and that depletion is inevitable in the foreseeable future. This is because economic, and more importantly energetic limits would prevent the extraction of any remaining oil. He recommends that the Arab economies diversify as quickly as possible by exploiting the many other mineral reserves that can be found in the region. Revenues from oil, while it lasts, may be used to accelerate this process. The author asserts that this exploitation of mineral reserves requires intense cooperation between the Arab states and even suggests that a Common Market of the Middle East be set up. This is a sobering look at what the end of oil will do to the producing nations—not just to the consuming ones.

239. Laherrère, J. H. "Reserve Growth: Technological Progress, or Bad Reporting and Bad Arithmetic?" *Geopolitics of Energy* 22, (April 1999).

Keywords: oil reserves, forecasting

The author suggests that oil reserve estimates and their upward revisions based on improvements in technology should be taken with a pinch of salt. The technology behind estimating reserves is among the worst in the oil industry and faulty reporting practices are rampant. Indeed, in many cases the oil reserve estimates should be revised downward. The author provides a lot of empirical evidence that justifies this suggestion. In many parts of the world the actual amount of recoverable oil in the fields was substantially below what was estimated, despite the application of the most advanced recovery technologies.

240. Manning, Robert A. *The Asian Energy Factor: Myths and Dilemmas of Energy, Security and the Pacific Future.* New York: Palgrave, 2000.

Keywords: oil reserves, alternative energy

There has been ongoing concern that industrialization efforts in Asia, especially in India and China are going to require ever increasing amounts of oil, thus hastening the demise of the age of hydrocarbons. Security concerns have also been raised since it was felt that much of the oil needed by these countries would necessarily have to be diverted from elsewhere, namely Europe and North America, leading to a possible armed conflict. However, the author of this book is cautiously optimistic about the energy future of these countries and does not foresee energy issues as being the cause of any conflict in the region. The development of new extraction and refining technologies will increase recoverable oil reserves, while at the same time technological developments in the end-use phase, such as the hyper-car, will lessen the demand for oil. Overall the author feels that, with proper governmental planning, the countries of this region can meet their development goals without energy becoming a bottleneck in their efforts.

241. Odum, Howard T., and Elisabeth C. Odum. *The Prosperous Way Down.* Boulder: University Press of Colorado, 2001.

Keywords: economics, environment, efficiency, policy

The Odums' concern in this book is premised on the fact that, at some time in the near future, resource scarcity and rising costs will cause the global economy to contract.

They are not concerned about the timing of this event as such, but much more with what can be expected and what our adapting strategies should be to conditions that force descent. The name given to this event is decession, the opposite of succession. Many people assume that the only way down is to crash and restart. However, according to the Odums, there are many systems that program orderly descent and decession that is followed later by growth and succession again. As an example they refer to the ecosystems and human cultures in northern latitudes in the past. These expanded and contracted seasonally; they decreased populations, stored information, and reduced function with such mechanisms as spore and seed formation, hibernation, migration, and staging inactivity and rest. The book goes on to give a comprehensive account of the changes that will accompany descent and warns that either we adapt with deliberate process or have these changes forced upon us with damaging repercussions. For each of these changes, the authors suggest strategies that could make descent less traumatic and more prosperous. In other words this is about "going down with style."

Of particular concern, from the perspective of energy, is a necessary decrease in population as well as a decrease in urban concentration since the extreme concentration of economic enterprises and people in cities is based on cheap fossil fuels. Luxury and wasteful uses of fuels must also be brought to a quick halt so that energy is channeled into more productive functions. There must be a declining dominance of automobiles and a global effort must be made to reforest the depleted lands of undeveloped countries to counteract the global greenhouse climate changes. Lower intensity agriculture must also be increased and diverse crop varieties that are more self-sustaining must also be restored. To continue the essentials of the world's civilization requires that global information networks be sustained. But this requires a priority in allocating electric power at a time when electric power from fossil and nuclear fuels will be extremely expensive. The authors conjecture that centers of civilization will reorganize around the foot of mountains with hydroelectric power. Many other policies regarding almost every dimension of human life round out the rest of this book.

242. Pimentel, David et al., "Renewable Energy: Economic and Environmental Issues." *BioScience.* 44, 8 (September 1994).

Keywords: solar, photovoltaics, oil reserves, alternative energy

The United States consumes approximately 25 percent of the world's fossil fuels with only 4.7 percent of the world's population. Currently it imports slightly more than half of its oil. Proven reserves of oil in the United States are expected to be exhausted by the year 2010 and it would be a major drain on the economy to import 100 percent of the oil. Gas reserves are expected to last a little longer, but not by much. Coal reserves, which are much more plentiful, cannot substitute for oil in all cases. Moreover, coal use contributes significantly to climate change. This paper evaluates several renewable energy technologies as well as what is possible with reference to conservation measures. It concludes that solar energy technologies, paired with energy conservation, have the potential to meet a large portion of future U.S. energy needs.

243. Price, David. "Energy and Human Evolution." *Population and Environment: A Journal of Interdisciplinary Studies* 16, 4 (March 1995).

Keywords: oil reserves, fuels

By using extrasomatic energy to modify more and more of its environment to suit human needs, the human population effectively expanded its resource base so that for long periods it has exceeded contemporary requirements. Human population was thus able to grow at an exponential rate, much like species that are introduced into extremely propitious environments, such as rabbits or cane toads in Australia or the Japanese beetle in the United States. However, the exhaustion of fossil fuels, which supply three-quarters of this extrasomatic energy, is not far off, and no other energy source is abundant and cheap enough to take their place. A collapse of the earth's human population cannot be more than a few years away. If there are survivors, they will not be able to carry on the cultural traditions of civilization, which require abundant, cheap energy. In the author's words, "It is unlikely, however, that the species itself can long persist without the energy whose exploitation is so much a part of its *modus vivendi.*"

244. Rauch, Jonathan. "The New Old Economy: Oil, Computers, and the Reinvention of the Earth." *Atlantic Monthly* 287, 1 (January 2001).

Keywords: oil reserves

This ebullient article had the misfortune to appear in the very month that California, Alberta, Ontario, and other states in North America felt an energy crunch and were faced with skyrocketing prices for fuels and electricity. The author claims that he was a "doom-sayer" for much of his life, always worried that an energy crunch was around the corner with disastrous consequences. Research into the "new old" economy however, has transformed him into a cornucopian. This economy combines the "old" resource-based economy with the "new" information-technology-based economy, giving new life, as it were, to the old economy. Nowhere is this better evident, according to the author, than in the oil industry where computers and associated technologies have revolutionized the finding and extracting of oil. Fast computers mean that geological structures can be mapped with stunning accuracy and wells can be drilled with remarkable precision so that fields can be exploited at rates that could only be dreamt of in the past. Three-dimensional seismic imaging technology allows geologists to virtually "walk around" structures that are thousands of feet deep or miles below the oceans. Horizontal drills guided by computers can now reach oil-bearing structures with greater accuracy as well. All of this means that the economic costs of searching for oil are far smaller since it is unlikely that drilling efforts will yield dry wells, as was the case in the past. According to the author these technologies have already allowed previously abandoned wells to be reopened to pump out significant amounts of oil that previously could not be mined. The upshot of all of this is that we need not fear an energy crunch since technology will keep us in plenty of fuels for the foreseeable future.

245. Ristiner, Robert A. and Jack J. Kraushaar. *Energy and the Environment.* New York: John Wiley & Sons, 1999.

Keywords: environmental, alternative energy, renewable energy

A comprehensive guide to the impact that different energy systems have on the environment. Most major concerns are addressed and a convincing argument is developed to demonstrate that many of our environmental problems have their origin in our quest for abundant and inexpensive energy. Chapters on renewable energy systems as well as on energy conservation add to the value of this book.

246. Saitoh, T. S. and T. Fujino. "Advanced Energy Efficient House with Solar Thermal, Photovoltaic, and Sky Radiation Energies." *Solar Energy* 70, 1 (2001).

Keywords: renewable, alternative energy, solar, photovoltaics

An energy independent residential house (Harbeman House: HARmony Between Man and Nature) incorporating sky radiation cooling, solar thermal and photovoltaic energies was built in Sendai, Japan in July 1996. This paper presents monitored results of this house since September 1996 to date. The detailed schematics of the house show how much is possible, with relatively little cost, in achieving energy efficiency through intelligent design efforts combined with the latest renewable energy technologies.

247. Schipper, L., ed. "On the Rebound: The Interaction of Energy Efficiency, Energy Use and Economic Activity." *Energy Policy* 28, 6-7 (June 2000).

Keywords: efficiency

The "rebound effect" has been identified by many experts to be a possible bottleneck in achieving sustainability in our use of energy. In its broadest sense, the rebound effect is simply the interaction of energy use with the efficiency of energy use: lower the energy required to do something and you will do a bit more of that thing. Case studies purporting to demonstrate this rebound effect are presented from Austria to India. The more theoretical articles suggest that data supporting the rebound effect are at best ambiguous.

248. Strodes, James. "Here We Go Again: The Oil Surplus Won't Last as Long as We Might Wish." *Barrons* (October 19, 1998).

Keywords: oil reserves, forecasting

A consensus has formed that within a few years new supplies of conventional oil energy will be outstripped by spiking world demand. Very soon after that the real volume of oil output will begin to shrink abruptly, even as demand growth coasts a bit higher. The author warns that the twenty-first century's supply disruptions and soaring prices will dwarf the OPEC crunches of 1973 and 1979. According to oil geologists, 90 percent of the globe's oil fields have already been tapped and many are already exhausted. Alternative energy technologies are poorly developed to take up the slack when the oil crunch comes.

249. Tainter, Joseph A. *Getting Down to Earth: Practical Applications of Ecological Economics.* New York: Island Press, 1996.

Keywords: fuels, economics, policy

As societies get more and more complex, the resolution of contemporary problems reaches a point of increasingly diminishing returns, requiring greater and greater inputs of effort (hence energy) to resolve these problems. If the capacity to make this effort is hindered or exhausted, there is an inevitable and catastrophic collapse. Tainter demonstrates this admirably by undertaking an analysis of the Roman Empire and its collapse along these very lines. Our own complex civilization is based on abundant, concentrated, high-quality energy from fossil fuels. With subsidies of inexpensive fossil fuels many negative consequences of industrialism effectively did not matter. Fossil fuels made industrialism, and all that flowed from it, such as science, transportation, medicine, employment, consumerism, high-technology war, a system of problem solving that was sustainable for several generations. Tainter says that if our efforts to understand and resolve such matters as global change involve increasing political, technological, economic and scientific complexity, as it seems they will, then the availability of energy per capita will be a constraining factor. To increase complexity on the basis of static or declining energy supplies would require lowering the standard of living throughout the world. In the author's words:

> In the absence of a clear crisis very few people would support this. To maintain political support for our current and future investments in complexity thus requires an increase in the effective per-capita supply of energy—either by increasing the physical availability of energy, or by technical, political, or economic innovations that lower the energy cost of our standard of living. Of course, to discover such innovations requires energy, which underscores the constraints in the energy-complexity relation.

There is a good discussion of possible paths to the future. There is the possibility that energy constraints will lead to a genuine collapse over a period of one or two generations with much violence and starvation and a great decline in population, at which point in time it is possible that cultural and economic simplicity will be adopted with much lower energy costs. The author considers the alternative of a "soft landing" rather than a crash through the voluntary adoption of solar technologies, green fuels, energy-conserving technologies, and less overall consumption as too utopian.

250.   Trainer, Ted. "The Death of the Oil Economy." *Earth Island Journal* (Spring 1997).

Keywords: oil reserves, forecasting

Economists argue that scarcity will result in price increases, making it more profitable to access poorer deposits. This, the author argues, is only plausible if one thinks only about dollar costs. The fact is, as an oil field ages, increasing amounts of energy must be exerted to pump the oil out. The cost of this energy must be subtracted from the total value of the energy in the oil retrieved. These two curves will intersect around the year 2005. Beyond that point, the energy required to find and extract a barrel of oil will exceed the energy contained in the barrel.

251.   Wiser, Wendell H. *Energy Resources: Occurrence, Production, Conversion, Use.* New York: Springer, 2000.

Keywords: fuels, generation, consumption

Designed as a textbook for students in a liberal education course and as a reference book for teachers of science in high schools, this comprehensive work does not flinch from a discussion of the wider environmental and security issues implicated in our use of energy. It will prove particularly useful to anyone with an interest in the subject of energy availability in the future. There are a lot of useful data on worldwide reserves of oil and natural gas as well.

252. Wohlgemuth, N., and F. Missfeldt. "The Kyoto Mechanisms and the Prospects for Renewable Energy Technologies." *Solar Energy* 69, 4 (2000).

Keywords: renewable, alternative energy, economics, policy

The Kyoto Protocol to the Climate Change Convention sets out legally binding emission targets and timetables for developed countries. In order to ease compliance, it allows countries to achieve their emission targets through the "Kyoto Mechanisms." These mechanisms comprise international emission trading (IET), joint implementation (JI), and a clean development mechanism (CDM). This paper analyses the capacity of the proposed mechanisms of the Kyoto Protocol to promote investment in renewable energy technologies (RETs). Analysis of abatement costing studies indicates that the increasing use of renewable energy tends to be a higher cost option compared with other greenhouse gas (GHG) technologies. This finding, however, does not make RETs unattractive for GHG mitigation as such because, apart from their vast technical potential to reduce GHG emissions, RETs have great capacity to contribute to other aspects of sustainable development.

253. World Oil's Editorial Advisors. "What's Ahead in 2001." *World Oil* (December 2000).

Keywords: oil reserves, policy, planning

Continuing high prices for oil are predicted for the near future with further increases down the road as demand for oil will continue to grow relative to supply, leading to shortages. The positive side of this picture is that it will encourage the bringing online of several technological innovations in oil recovery that low oil prices had kept offline. It is hoped that the use of these technologies and the added incentive for greater exploration will help to offset some of the shortages, because no matter what, oil and gas are still going to be dominant energy sources for decades to come.

254. Youngquist, Walter. *Mineral Resources and the Destinies of Nations.* Portland, Ore.: National Book Company, 1990.

Keywords: oil reserves

This book looks at the impact of the present, huge demand for minerals—both metal and energy—in the industrialized and the developing world. The author brings a wealth of background information to accomplish this task and reaches into the domains of national and international economics, politics, technology, demography, history, engineering, and geology. Oil, an energy mineral that is so inextricably meshed with the global economy and world politics receives special treatment in this absorbing and fascinating account. Youngquist considers several scenarios of what will happen after the end of the age of oil. None of the scenarios is pleasant, because there are no complete substitutes for oil's versatility. He warns us that the easy technologies have already been developed and the easily exploitable mineral resources are now rapidly being used up. We are coming to the end of one era and about to enter another.

This is not just a record of minerals in world events until 1990, but also an attempt to make people aware of the past, present, and future importance of these resources. Much of written history tends to ignore these fundamentals, perhaps because past historians generally were not educated in the realm of natural resources but were oriented toward social and political matters and did not recognize the very basic importance of the energy and mineral foundations of society. The author goes on to hope "that historians of the present and the future will see this more closely, as the energy and mineral demands of the industrialized world soar, and conflicts over them result—the Persian Gulf oil problems of the 1980s being an example."

255. Youngquist, Walter. *GeoDestinies: The Inevitable Control of Earth Resources Over Nations and Individuals.* Portland, Ore.: National Book Company, 1997.

Keywords: oil reserves

This remarkable and comprehensive work extends and completes what the author had started in his previous work, *Mineral Resources and the Destinies of Nations,* (*see* no. 254). No aspect of the multifaceted role minerals play in our lives, and in the lives of nations, has been left unanalyzed. There is a lively discussion of the historical role of minerals as well as future prospects as the appetites of the developed and developing world become even more voracious with each passing day and as the resources themselves get to be scarcer. While this book is a must for anyone who wants to find out more about the wider significance of mineral resources, chapter 27, "Myths and Realities of Mineral Resources," is required reading for anyone who wants to get beyond the many myths that surround minerals in general and oil in particular. This is where Youngquist is at his most incisive in this always lucid book, and provides the reader a great service by debunking a whole host of myths—and the reasons why they have arisen—around minerals.

From the perspective of this bibliography, some of the most persistent of these myths surround oil, such as the common perception that there is no oil supply problem for the United States. Youngquist quickly dispels this myth by showing that in fact the United States passed the point of oil self-sufficiency in 1970, and has been an importer of oil ever since. In fact, it is now importing much more oil than it produces. Another persistent myth that is dispelled is that drilling deeper will lead to the discovery of more oil. The author correctly reveals that in fact oil occurs in sedimentary rocks which are a fairly thin part of the Earth's crust and that 16,000 feet is, with a few exceptions, the limit of oil occurrence. Below that depth, because of the temperature of the Earth, only gas exists. Youngquist also tackles the issue of whether alternative energy resources can readily replace oil, by posing this statement as a myth as well. Alternative energy sources can replace oil in its energy uses, but in some uses much less conveniently than in others. For instance, replacing gasoline, kerosene, and diesel fuel for use in vehicles, particularly airplanes, will be much more difficult. Moreover, it is naïve to think that alternative energy sources can simply be plugged into our present economic system and lifestyle, and things will go on as usual. The limits to energy conservation, especially in the face of a growing population are also well described.

256. Youngquist, Walter. "Spending Our Great Inheritance—Then What?" *Geotimes* (July, 1998).

Keywords: oil reserves, forecasting, alternative energy

When the future of oil is discussed, the common question asked is: "How long will it last?" This is the wrong question. Insignificant amounts of oil will probably be produced in the year 2100 and perhaps beyond. The critical date is when the peak of oil production is reached and the world's demands can no longer be supplied. From then on, there will be less and less oil to divide, in contrast to the current (1998) happy situation where we have more and more to divide. It is probable that the decline of world oil production will affect more people in more ways than any other event in human history. The author explores several scenarios regarding oil production and concludes that world oil production will peak and then decline within the first two decades of the twenty-first century, no matter how optimistic one's outlook is. A close analysis of alternative energy resources and technologies is explored, and there is a slim chance that in the future many of these would be able to substitute for oil. However, even if we assume that alternative sources could somehow fill the gap left by the exhaustion of oil supplies, it will take an unacceptably long time to put them into production. The author concludes that while all alternative energy technologies must be drawn upon, oil will still be sorely missed, and most of us alive today will probably live to see that day.

AUTHOR INDEX

Note: This index is to annotation numbers.

KEYWORD INDEX

Note: This index is to annotation numbers.

ABOUT THE AUTHORS

Willem H. Vanderburg is the founding director of the Centre for Technology and Social Development in the faculty of applied science and engineering at the University of Toronto and holds cross-appointments in the Institute for Environmental Studies and the department of sociology. He is the editor-in-chief of the *Bulletin of Science, Technology & Society,* and the author of *Perspectives on Our Age: Jacques Ellul Speaks of His Life and Work* (House of Anansi Press, 1997), *The Growth of Minds and Cultures* (University of Toronto Press, 1985), and *The Labyrinth of Technology* (University of Toronto Press, 2000).

Namir Khan is a lecturer at the Centre for Technology and Social Development in the faculty of applied science and engineering at the University of Toronto. He is the managing editor of the *Bulletin of Science, Technology & Society.*